ACTION RESEARCH
THIRD EDITION

Ernest T. Stringer

Curtin University of Technology, Australia

SAGE Publications
Los Angeles • London • New Delhi • Singapore

For information:

Sage Publications, Inc.
2455 Teller Road
Thousand Oaks, California 91320
E-mail: order@sagepub.com

Sage Publications Ltd.
1 Oliver's Yard
55 City Road
London EC1Y 1SP

Sage Publications India Pvt Ltd
B 1/I 1 Mohan Cooperative Industrial Area
Mathura Road, New Delhi 110 044
India

Sage Publications Asia-Pacific Pte Ltd
33 Pekin Street #02-01
Far East Square
Singapore 048763

Printed in the United States of America

Library of Congress Cataloging-in-Publication Data

Stringer, Ernest T.
Action research / Ernest T. Stringer. —3rd ed.
p. cm.
Includes bibliographical references and index.
ISBN-13: 978-1-4129-5222-4 (cloth: alk. paper)
ISBN-13: 978-1-4129-5223-1 (pbk.: alk. paper)
1. Human services—Research. 2. Action research. I. Title.
HV11.S835 2007
361—dc22

2006034022

Printed on acid-free paper

07 08 09 10 11 10 9 8 7 6 5 4 3 2 1

Acquiring Editor: Lisa Cuevas Shaw
Editorial Assistant: Karen Margrethe Greene
Production Editor: Denise Santoyo
Copy Editor: Mary L. Tederstrom
Typesetter: C&M Digitals (P) Ltd.
Marketing Manager: Stephanie Adams

CONTENTS

FOREWORD

I was recently asked, in another context, to speculate on the future of educational research in the event it was to change in a direction that I hoped it would. My response was that I hoped *all* research that might properly be called *human* inquiry would exhibit three characteristics: decentralization, deregulation, and cooperativeness in execution. I believe that this book describes a mode of inquiry that fits all of these specifications.

By *decentralization* I meant to indicate a movement away from efforts to uncover generalizable "truths" toward a new emphasis on local context. The hiatus between theory and practice has been remarked on too often to require repetition here. The reason for that hiatus, I have long asserted, lies in the gap between general laws and specific applications; such laws can have, at best, only probabilistic implications for specific cases. The fact, for example, that 80% of patients presenting a given set of symptoms are likely to have lung cancer does not imply that a particular patient with those symptoms ought to immediately be rushed into surgery.

We have witnessed, over the past half century or so, determined efforts to find general solutions to social problems, be they low pupil achievement, drug abuse, alcoholism, AIDS, or other challenges. The cost to national economies has been prodigious, and there is precious little to show for it, little "bang for the buck," as some folks are wont to say. It ought to be apparent by now that generalized, one-size-fits-all solutions do not work. The devil (or God, if you prefer) is in the details. Without intimate knowledge of local context, one cannot hope to devise solutions to local problems. *All* problems are de facto local; inquiry must be decentralized to the local context.

By *deregulation* I meant to indicate a movement away from the restrictive conventional rules of the research game, the overweening concern with validity, reliability, objectivity, and generalizability. I have argued elsewhere that these methodological criteria can have meaning only within a paradigm of inquiry that is defined in the conventional way and, specifically, based on the premise of concrete, tangible reality. However useful the premise of such a reality may be in the physical sciences (and history has shown it to be useful there indeed), it is simply irrelevant in the arena of human inquiry, for in that arena there is no tangible reality; everything that social inquirers study depends on mental constructions and mental interpretations. Thus the usual distinction between ontology (the nature of reality) and epistemology (how one comes to know that reality) collapses: Inquirers do not "discover" knowledge by watching nature do its thing from behind a one-way mirror; rather, it is literally created by the interaction of inquirers with the "object" (construct) into which they have inquired. Whatever may be the criteria for research quality in this new arena, the conventional criteria clearly do not fit.

By *cooperativeness in execution* I meant to indicate a style of inquiry in which there is no functional distinction between the researcher and the researched "subjects" (in conventional parlance). They are all defined as participants, and they all have equal footing in determining what questions will be asked, what information will be analyzed, and how conclusions and courses of action will be determined. These participants, sometimes called *stakeholders* or *local members*, may include some with special training in inquiry, but if so, these specialists have no privilege in determining how the study will go; *all* participants share the perquisites of privilege. I insist on this joint approach both because local stakeholders are the only extant experts on local culture, beliefs, and practices and because moral considerations require that local perspectives be honored.

The reader will quickly see, I am sure, that this book conforms to these stipulations exceptionally well. On the matter of decentralization, Dr. Stringer takes issue with "applied scientific expertise" aimed at "eradicat[ing] [a] problem by applying some intervention at an individual or programmatic level" (Chapter 1). He argues that there is evidence to suggest that "centralized policies and programs generated by 'experts' have limited success" in overcoming problems. His proposal for *community*-based action research returns the focus of inquiry to the local context.

On the matter of deregulation, Dr. Stringer asserts, in a previous edition, that "formal research operates at a distance from the everyday lives of practitioners, and largely fails to penetrate the experienced reality of their day-to-day work. The objective and generalizable knowledge embodied in social and behavioral research often is irrelevant to the conflicts [they] encounter"

(Chapter 1). He opts instead for the resurgence of action research, which is fundamentally different from the classic approach of defining variables and generating, through hypotheses and tests, explanations for why people behave as they do. That action research may not conform to conventional criteria of research rigor is much less important than the fact that it takes a more democratic, empowering, and humanizing approach; assists locals in extending their own understanding of their situations; and helps them to resolve the problems they see as important.

On the matter of cooperativeness in execution, Dr. Stringer calls for a form of inquiry that represents a "moral intertwining" of all participants, including, of course, the inquirer. The major concerns that occupy the attention of all but the most ardently conservative investigators—empowerment, democracy, equity, liberation, freedom from oppression, and life enhancement—are central to community-based action research. Ethics and morality are inscribed as essential features of human inquiry—not simply as standards to be met in the interest of humanity, but as standards that determine the very nature of study outcomes. Values cannot be separated from the core of an inquiry by the simple expedient of claiming objectivity, for findings are literally created by the inquiry process. And that process is permeated by values at every step.

Now, all of this has a very theoretical sound, and that lack of grounding is one of the major difficulties that has accompanied various calls for new approaches and new paradigms. However important theoretical justifications may be, they have little value if their implications are not translated into forms useful to the *practitioners* of this nontraditional social research. It is this urgent need to which this book responds, and, in my opinion, responds very well. Let me mention several of its features that commend it for this purpose:

- The language of the book is eminently accessible to practitioners who may be unfamiliar with typical research parlance. This is clearly not a book written solely for professional researchers; it is within the grasp of any reasonably literate reader. There is no arcane language to confuse the unwary.

- Every procedure described is accompanied by step-by-step instructions. The professional researcher may find these instructions unnecessarily detailed, but the novice will surely appreciate being helped at each juncture. And even the professional will find the details useful precisely because the approach differs so dramatically from what is normally found in a research methods text.

• The book is written to be useful to a wide variety of audiences, including but not limited to teachers, health workers, social workers, community workers, counselors, and many other lay workers and professionals. But to say that a book is intended to be useful for this or that audience has little meaning unless those audience members can find themselves represented in examples and depicted in situations. This aim is well met in this volume.

• The representations made by the author are illustrated throughout by personal anecdotes. Indeed, the inclusion of these anecdotes is perhaps the most useful feature of the book, because it makes clear to the reader that the author is speaking from a wealth of personal experience. He has not only "talked the talk" but also "walked the walk." The reader can have confidence that every assertion has been validated in a real-life situation and that every procedure recommended has been found to work in some real-world instance. And precisely because the author makes plain that his background is not essentially dissimilar from the reader's, the reader will find these anecdotes confidence building. What is being proposed is not only possible for an expert but also possible for an everyday reader. And if it can work with a cultural group as different from "typical" Westerners as are Australian Aboriginals, it can probably work in just about any cultural setting.

It is not only the practitioner of community-based action research with whom Dr. Stringer is concerned, however; he is equally interested in persuading the more conservative academic of the soundness of his proposals. Accordingly, he includes a final chapter that sets the whole in a proper theoretical context. His reflections address the question of the legitimacy of his proposed approach in order to secure for it a wider acceptance, discuss the issues of power and control, and set action research in the context of the postmodern position. The skeptic and the critic may or may not find his arguments persuasive, but, if taken seriously, these arguments raise important questions about which each reader will have to satisfy himself or herself. And the novice may take comfort in the fact that rational arguments do exist for the practice of a kind of inquiry that many practitioners have intuitively felt to be right (for them) but about which they have felt insecure on the grounds of rigor or objectivity. There is indeed more in heaven and earth than has been dreamed of in the received philosophy.

—*Egon G. Guba*

PREFACE

THE PURPOSE OF THIS BOOK

This book has been written for those workers, both professional and nonprofessional, who provide services to people in community, organizational, or institutional contexts. It speaks, therefore, to teachers, health professionals, social workers, community and youth workers, businesspeople, planners, and a whole range of other people who function in teaching, service delivery, or managerial roles. Its purpose is to provide a set of research tools that will enable people to deal effectively with many of the problems that confront them as they perform their work. I attempt to provide clear guidelines to enable novice practitioner researchers to move comfortably through a process of inquiry that provides effective solutions to significant problems in their work lives. The "biographical bulletins" that punctuate the text are designed to clarify meaning and to increase understanding of relevant facets of research processes.

THE ROLE OF THE RESEARCHER

A common approach to action research envisages processes of inquiry that are based on a practitioner's reflections on his or her professional practices. It has been clear to me for many years that when practitioners remain locked into their own perceptions and interpretations of the situation, they fail to take into account the varied worldviews and life experiences of the people with whom they work. They fail to understand the fundamental dynamics that determine the way their clients, students, patients, or customers will behave in any given circumstance. Community-based action research works from the assumption that all people affected by or having an effect on an issue should be involved in the processes of inquiry.

In these circumstances the task of the practitioner researcher is to provide leadership and direction to other participants or stakeholders in the research process. I therefore speak throughout this book to those who coordinate or facilitate the research as *research facilitators.* For ease and clarity, I often shorten this to *researchers* or *facilitators.* Practitioners may accept the role of researcher, therefore, when they enact research processes with groups of other stakeholders—students, clients, administrators, and so on. Ultimately, however, all participants in the research process should rightfully be called researchers insofar as they engage in deliberate processes of inquiry or investigation with the intent of extending their understanding of a situation or a problem.

In many situations, the demands of professional or community life prevent practitioners from taking such an active leadership role, and these practitioners may call on the services of outside consultants to perform coordinating, facilitating functions. In such cases, the consultants would accept the role designated in this book as researchers or research facilitators.

THIRD EDITION

The first and second editions of this text were largely derived from my experience in a variety of Australian contexts. Much of my initial writing was based on work with Aboriginal community groups and organizations and with government departments, business, and industry. Since that time I have had opportunities to extend my experience by working in a number of parts of the United States and in East Timor. In these circumstances I have been able to verify my faith in the processes incorporated into this book, engaging in projects with Hispanic, African American, Anglo-American, and a range of social groups that comprise the people in these contexts.

I have attempted to integrate the flavor of some of this work into this text, but limitations of space prevent a full exploration of my experience. Suffice it to say that I continue to find the process of unleashing the energy and potential of the people to be both humbling and fulfilling. I have watched in admiration as a small neighborhood group in New Mexico engaged in a delightfully effective action research project in their local school and have been filled with awe at the emerging power of the people of East Timor as they rebuilt a school system devastated by departing occupying forces.

So it is "ordinary people" who give me the most satisfaction, as I watch them grow in skill and power as they accomplish wonderful outcomes through the systematic application of the research processes outlined in this book. But I am also impressed with people in positions of power and authority who, given the opportunity, provide the support and resources to accomplish these ends. Recently I wrote a small biography for an action research association in which I described the meaning that action research has given to my experience:

> Action research is more than a process, to me. It's a passion. . . . I have been constantly reassured that the wisdom of the people, and the knowledge they have of their own situation provides a much better basis for action than ideas that come from my own experience. I have been gratified by the deeply purposeful work in which people have engaged, delighted in the very practical, immediate outcomes they achieve, and heartened by the sense of empowerment that comes to them in the process. It is the energy and enthusiasm that results from these participatory processes that continues to inspire me.

My hope for you as you read this edition is that you will also come to be inspired, heartened, and gratified through engaging the potential of the people with whom you work through the systematic and soulful application of action research.

STRUCTURE OF THE BOOK

This book provides a resource for practitioners, to assist them in their efforts to conduct inquiry and to hone their investigative skills so that they will formulate effective solutions to the deep-rooted problems that detract from the quality of their professional lives. In the chapters that follow, I present an approach to inquiry that helps practitioners to explore systematically the real-life problems they experience in their work contexts and to formulate effective and sustainable solutions that enhance the lives of the people they serve.

Chapter 1 reviews the nature of research and provides an overview of community-based action research. It includes a discussion of the basic values inherent in this approach to research and suggests professional, organizational, and community contexts where action research might be appropriately applied. The chapter also presents a basic routine—look, think, act—that serves as a framework to guide the research process. It presents a clear description of the role of the researcher and introduces the reader to the now copious literature on action research.

Chapter 2 presents the theoretical foundations of action research, assisting the reader in understanding why it is carried out in the manner suggested in these pages. The chapter also presents the principles of action research that are derived from both the values inherent in the methodology and the pragmatic forces that shape human activity.

Chapter 3 presents processes for constructing an effective action research plan. It includes the preliminary activity required to ensure that an adequate sample of people is included in the research and processes for establishing initial contact with them. The chapter also maps out procedures for ensuring that research is carried out ethically and rigorously.

Chapter 4 focuses on data gathering—the different ways of gathering information that assist research participants in extending their understanding of the issue they investigate. The chapter provides techniques for gathering and recording data—observation, interviews, surveys, document searches, and so on.

Chapter 5 provides a description of the process through which stakeholders interpret or analyze the data. It presents two major processes for distilling qualitative information to identify key features and elements that enable participants to develop insightful understandings of the issue investigated.

Chapter 6 presents procedures that enable participants to formulate practical solutions to their problems. Three phases of activity are described: (a) planning, in which priorities are set and tasks defined; (b) implementing, the supporting, modeling, and linking activity that enables participants to accomplish their tasks; and (c) evaluating, through which participants review their progress.

Chapter 7 focuses on more complex processes for developing sustainable solutions to the deep-seated problems often contained within large organizations, government agencies, business corporations, and community contexts. This chapter provides an orientation to strategic planning processes and organizational arrangements for implementing activities and evaluating outcomes.

Chapter 8 suggests ways of organizing and formulating formal reports. It contrasts ways of organizing experimental scientific reports with structures appropriate for interpretive action research and then presents a more detailed approach to formulation of the latter. Although this style of reporting is relevant to formal reports required in bureaucratic settings and funding agencies, it is particularly suited to academic reports required for university theses and dissertations.

Chapter 9 reviews and reflects on the research processes described in the preceding chapters. Here I address the question of the legitimacy of community-based action research and provide a discussion of how I make sense of this approach to inquiry with reference to the perspectives of postmodern social theory.

Appendixes provide a listing of the voluminous action research resources now available on the Internet and examples of reports of two action research projects.

ACKNOWLEDGMENTS

In previous editions of this book I have acknowledged many of the people who have contributed to the development and writing of the material encompassed by this book. My debt to them continues, and the number of people who continue to participate in the development of ideas likewise proliferates. My first forays into the research literature, initiated by my good friend Geoff Mills, eventually moved me to write the first edition of this volume under the wonderful guidance of Egon Guba. Six books later, I am still refining my understanding of action research and find myself both nurtured and challenged by my interactions with colleagues, friends, students, and my family. Similarly, I find that my contributions as a member of the editorial board of the *Action Research Journal* and as president of the Action Learning, Action Research and Process Management Association extend my understanding and my capacity to engage in effective action research. I thank all those with whom I have interacted in these contexts, as they provide the rich diversity of relationship and perspective that enriches my professional and community life.

Centrally, though, I acknowledge the strongly supportive role provided by my partner, wife, companion, and friend, Rosalie Dwyer, who continues to nurture the creative spirit within me. Her thoughtful comments, active encouragement, detailed scrutiny, and constant companionship provided the personal and spiritual energy that sustained me through the process of writing and editing.

Finally, I would like to acknowledge the fine work of the editorial team at Sage who continue to provide the detailed support required to engage the extended processes of writing, editing, and production of this book. Thank you most especially to Lisa Cuevas Shaw and Karen Greene in editorial, Denise Santoyo in production, copy editor Mary Tederstrom, and Stephanie Adams in marketing.

ONE

RESEARCH IN PROFESSIONAL AND PUBLIC LIFE

THE PURPOSES AND APPLICATIONS OF ACTION RESEARCH: WHO DOES ACTION RESEARCH, AND WHY DO THEY DO IT?

Action research is a systematic approach to investigation that enables people to find effective solutions to problems they confront in their everyday lives. Unlike traditional experimental/scientific research that looks for generalizable explanations that might be applied to all contexts, action research focuses on specific situations and localized solutions. Action research provides the means by which people in schools, business and community organizations; teachers; and health and human services may increase the effectiveness of the work in which they are engaged. It assists them in working through the sometimes puzzling complexity of the issues they confront to make their work more meaningful and fulfilling.

For many people, professional and service occupations—teaching, social work, health care, psychology, youth work, and so on—provide appealing avenues of employment. These occupations have the potential to provide them with meaningful and fulfilling work that they find intrinsically rewarding. Increasingly, however, people in these sectors find their work to be more demanding and less satisfying. They often struggle to balance growing demands on their time and energy as their workloads continue to expand, and they are routinely confronted by problems rarely encountered 20 or 30 years ago.

The pressures experienced in professional practice reflect tensions that exist in modern society. The complex influences that impinge on people's everyday social lives provide a fertile seedbed for a proliferating host of family, community, and institutional problems. Professional practitioners and agency workers are increasingly held accountable for solutions to problems that have their roots in the deeply complex interaction between the experiences of individual people and the realities of their social lives: stress, unemployment, family breakdown, alienation, behavioral problems, violence, poverty, discrimination, conflict, and so forth.

Although adequately prepared to deal with the technical requirements of their daily work, practitioners often face recurrent crises that are outside the scope of their professional expertise. Teachers face children disturbed by conflict in their homes and communities, youth workers encounter resentful and alienated teenagers, health workers confront people apparently unconcerned about life-threatening lifestyles and social habits, and social and welfare workers are strained past their capacity to deal with the impossible caseloads spawned by increasing poverty and alienation.

There is an expectation in social life that trained professionals, applying scientifically derived expertise, will provide answers to the proliferating problems that confront people in their personal and public lives. Community responses to crises that arise from drug abuse, crime, violence, school absenteeism, and so on invariably revolve around the use of a social worker, youth worker, counselor, or similar type of service provider whose task it is to eradicate the problem by applying some intervention at an individual or programmatic level. These responses have failed to diminish the growing social problems that have multiplied much faster than the human and financial resources available to deal with them. Moreover, evidence suggests that centralized policies and programs generated by "experts" have limited success in resolving these problems. The billions of dollars invested in social programs have failed to stem the tide of alienation and disaffection that characterize social life in modern industrial nations.

If there are answers to these proliferating social problems, it is likely that centralized policies will need to be complemented by the creative action of those who are closest to their sources—the service professionals, agency workers, students, clients, communities, and families who face the issues on a daily basis. Centralized policies, programs, and services, I suggest, should allow practitioners to engage the human potential of all people who contribute to the complex dynamics of the contexts in which they work. Policies and programs should not dictate specific actions and procedures but instead should provide the resources to enable effective action that is appropriate to particular places. The daily work of practitioners often provides many opportunities for them to acquire valuable insights into people's social worlds and to assist them in formulating effective solutions to problems that permeate their lives.

We therefore need to change our vision of service professionals and administrators from mechanic/technician to facilitator and creative investigator. This new vision rejects the mindless application of standardized practices across all settings and contexts and instead advocates the use of contextually relevant procedures formulated by inquiring and resourceful practitioners. The pages that follow describe some of the ways professional and community workers can hone their investigative skills, engage in systematic approaches to inquiry, and formulate effective and sustainable solutions to the deep-rooted problems that diminish the quality of professional life. This volume presents an approach to inquiry that seeks not only to enrich professional practice but also to enhance the lives of those involved.

As a young teacher, I had the rare experience of being transferred from the relative security of a suburban classroom to a primary school in a remote desert region of Western Australia. My task was to provide education for the children of the traditional hunter-gatherer Aboriginal people who lived in that area. On my first day in class, I was confronted by a wall of silence that effectively prevented any possibility of teaching. The children refused to respond verbally to any of my queries or comments, hanging their heads, averting their eyes, and sometimes responding so softly that I was unable to hear what they said. In these discomfiting circumstances, I was unable to work through any of the customary routines and activities that had constituted my professional repertoire in the city. Lessons were abbreviated, avuncular, and disjointed, and my professional pride took a distinct jolt as an ineffective reading lesson followed an inarticulate math period, preceding the monotony of my singular voice through social studies.

The silence of the children in the classroom was in marked contrast to their happy chatter as we walked through the surrounding bush in the afternoon, my failing spirits leading me to present an impromptu natural science lesson. I was eventually able to resolve many of the problems that faced me in this unique educational environment, but the experience endowed me with an inquiring professional mind. In these circumstances, all the taken-for-granted assumptions of my professional life rang hollow as I struggled to understand the nature of the problems that confronted me and to formulate appropriate educational experiences for this wonderfully unique group of students. Texts, curricula, teaching materials, learning activities, classroom organization, speech, interactional styles, and all other facets of classroom life became subjects of inquiry and investigation as I sought to resolve the constant stream of issues and problems that emerged in this environment. To be an effective teacher, I discovered that it was necessary to modify and adapt my regular professional routines and practices to fit the children's cultural realities.

The legacy of that experience has remained with me. Although I have long since left school classrooms behind, the lessons I learned there still pervade all my work. I engage all professional, organizational, and community contexts with a deep sense of my need to explore and understand the situation. Processes of inquiry enable me to engage, examine, explore, formulate answers, and devise responses to deal effectively with the issues before me.

In these situations, I now cast myself as a research facilitator, working with and supporting people to engage in systematic investigation that leads to clarity and understanding for us all and to provide the basis for effective action. In many places in the United States, Canada, Australia, East Timor, and Singapore I use techniques and procedures that can be fruitfully applied to the day-to-day work of people in schools, organizations, and community settings. I am now a practitioner-researcher.

RESEARCH: METHODICAL PROCESSES OF INQUIRY

Research is systematic and rigorous inquiry or investigation that enables people to understand the nature of problematic events or phenomena. Research can be characterized by the following:

- A *problem or issue* to be investigated
- A *process* of inquiry
- *Explanations* that enable individuals to understand the nature of the problem

Research can be visualized as nothing more than a natural extension of the activities in which we engage every day of our lives. Even for simple problems—Where are my blue socks? Why did the cake burn?—we ask questions that enable us to analyze the situation more carefully. (I wore my blue socks yesterday; I probably put them with the laundry. Perhaps I overheated the oven, or maybe I left the cake in the oven longer than I should). Tentative analysis enables us to understand the nature of the problem and to work toward a potential solution. (I looked in the laundry, and the socks were there. Next time I baked a cake, I lowered the temperature of the oven and did not burn the cake.)

Formal research is an extension of these day-to-day inquiries. The success of scientific research can be ascribed to its insistence on precise and rigorous formulation of description, observation, and explanation. The meticulous association of what is observed and what is explained provides explanations whose power and efficacy enable us to predict and control many facets of the physical world. The outcomes of scientific research are embodied in the technical achievements that continue to transform our modern world. The miracles of construction, manufacture, communication, and transport that have now entered the daily lives of those living in wealthy nations are testament to the huge advances in knowledge that have resulted from science.

Less successful, however, have been the attempts of the social and behavioral sciences to emulate the accomplishments of the physical sciences. Despite a profusion of theory, the application of scientific method to human events has failed to provide a means for predicting and controlling individual or social behavior. Teachers, health workers, and human service practitioners often find that the theoretical knowledge of the academic world has limited relevance to the exacting demands of their everyday professional lives. The objective and generalizable knowledge embodied in social and behavioral research often is only marginally relevant to the situations they encounter in their daily lives and has little application to the difficulties they face.

Action research, however, is based on the proposition that generalized solutions may not fit particular contexts or groups of people and that the purpose of inquiry is to find an appropriate solution for the particular dynamics at work in a local situation. A lesson plan, a care plan, or a self-management plan that fits the lifeworld of a middle-class suburban client group may be only tangentially relevant in poor rural or urban environments or to people whose cultural lives differ significantly from the people who serve them. Generalized solutions must be modified and adapted in order to fit the context in which they are used.

The wheel provides a good metaphor to understand the nature of this process. Wheels provide a general solution to the problem of transporting objects from one place to another though there are many different purposes to which they are put. Consider the different purposes, parameters, and processes required to use wheels for the following objects:

- A jumbo jet
- A small, single-engine aircraft
- A truck

- A child's tricycle
- A skateboard

Although the general concept of the wheel applies to all, there is considerable difference in the form wheels must take to enable them to achieve the general purpose that underlies their "wheel" function. Careful and systematic design work is required to ensure that the "wheel" functions efficiently and safely for the particular context in which it operates.

The same applies in most fields of human activity. Although there are general processes involved in, for instance, teaching, health care, social work, business, and industry, there is always a need to modify and adapt those processes for the particular people involved and the place where they are applied. Action research provides the means to systematically investigate issues in diverse contexts and to discover effective and efficient applications of more generalized practices. The primary purpose of action research is to provide the means for people to engage in systematic inquiry and investigation to "design" an appropriate way of accomplishing a desired goal and to evaluate its effectiveness.

As will become evident, however, the practitioner does not engage this work in isolation. An assumption of action research is that those who have previously been designated as *subjects* should participate directly in the research. Community-based action research works on the assumption that all people who affect or are affected by the issue investigated should be included in the processes of inquiry. The "community" is not a neighborhood or a suburb, but a community of interest. Action research is a participatory process that involves all those who have a stake in the issue engaging in systematic inquiry into the issue to be investigated. Professional practitioners, as research facilitators, engage their communities of interest in careful and systematic explorations that provide them with knowledge and understanding that, in very direct ways, improve the quality of their lives.

As a graduate student, I was excited by the possibilities of the hypothetico-deductive method of research. Here, I thought, was the means for obtaining answers to the significant social problems that concerned me. By careful measurement of critical variables related to the problem under investigation and precise definition and measurement of the relationship between them, it would be possible to describe the genesis of the problem and take appropriate steps to resolve it. The major task was simply to identify the appropriate variables, measure them, and analyze them using appropriate statistical techniques. It would then be possible to predict the ways people would behave or perform in particular circumstances and to take remedial action.

In my attempts to understand the reason for low achievement levels of minority students, I set out to map the variables that had impacts on their school performance, with the intent of first defining, then measuring, the extent of the relationships between the factors that related to their academic performance. I equipped myself for the task by taking many courses in descriptive and inferential statistics and experimental and survey research methods, while concomitantly immersing myself in the voluminous and burgeoning research literature that spoke to these issues.

My disenchantment with this approach to inquiry came through courses in anthropology and sociology, which awakened me to a relativistic social universe. It was a perspective that revealed a world far different from the mechanistic, soulless vision that, at that time, was favored by scientists. For me, the new paradigm encompassed social spheres composed of the changing lives of people as they created and re-created their realities according to systems of meaning inherent in their differing situations. The social world, I discovered, was not static and mechanistic but dynamic and changing, encapsulated by and redefined continuously by the symbolic systems of thought and language through which human beings fashion their physical and social universe.

The transformation of my thought was dramatic. I realized that I, as an impartial, objective observer, could never hope to define, discover, or measure the worlds of meaning that embodied human behavior in any social setting; that any hypothesis or explanation that I formulated at a distance from those worlds of meanings could bear little meaningful relationship to the actions and activities of the people who inhabited them; and that any interpretation of their behavior that failed to take into account the ways in which participants defined and described their situations must necessarily fail as an explanatory system.

Through rigorous exploration and inquiry, I had acquired a new vision of the world, a new way of comprehending the complexity that surrounded me and, in doing so, reconfigured my relationship with the people I proposed to study. Far from defining and describing the variables that explained the nature of their existence and generating explanations about why they behaved as they did, I was now cast in a position of ignorance in relation to those who had previously been potential subjects of study. Only through them, the cultural experts in their own settings, could I acquire the information that would enable me to understand how they behaved as they did. A new "researcher" emerged.

A Basic Routine

Action research is a collaborative approach to *inquiry* or *investigation* that provides people with the means to take systematic *action* to resolve specific problems. Action research is not a panacea for all ills and does not resolve all problems but provides a means for people to "get a handle" on their situations and formulate effective solutions to problems they face in their public and professional lives. The basic action research routine provides a simple yet powerful framework—look, think, act (see Box 1.1)—that enables people to commence their inquiries in a straightforward manner and build greater detail into procedures as the complexity of issues increases. The terms in parentheses in Box 1.1 show how the phases of the routine relate to traditional research practices.

Box 1.1 A Basic Action Research Routine

Look	• Gather relevant information (Gather data) • Build a picture: Describe the situation (Define and describe)
Think	• Explore and analyze: What is happening here? (Analyze) • Interpret and explain: How/why are things as they are? (Theorize)
Act	• Plan (Report) • Implement • Evaluate

The "look, think, act" routine is but one of a number of ways in which action research is envisaged. Kemmis and McTaggart (1999), for instance, present action research as a spiral of activity: plan, act, observe, reflect. Different formulations of action research reflect the diverse ways in which the same set of activities may be described, although the processes they delineate are similar. There are, after all, many ways of cutting a cake.

Although the "look, think, act" routine is presented in a linear format throughout this book, it should be read as a continually recycling set of

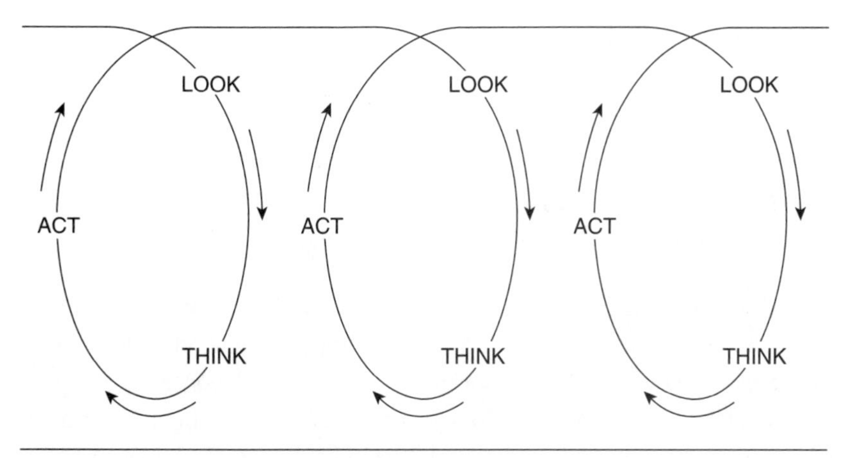

Figure 1.1 Action Research Interacting Spiral

activities (see Figure 1.1). As participants work through each of the major stages, they will explore the details of their activities through a constant process of observation, reflection, and action. At the completion of each set of activities, they will review (look again), reflect (reanalyze), and re-act (modify their actions). As experience will show, action research is not a neat, orderly activity that allows participants to proceed step-by-step to the end of the process. People will find themselves working backward through the routines, repeating processes, revising procedures, rethinking interpretations, leapfrogging steps or stages, and sometimes making radical changes in direction.

In practice, therefore, action research can be a complex process. The routines presented in this book, however, can be visualized as a road map that provides guidance to those who follow this less traveled way. Although there may be many routes to a destination, and although destinations may change, travelers on the journey will be able to maintain a clear idea of their location and the direction in which they are heading.

The procedures that follow are likely to be ineffective, however, unless enacted in ways that take into account the social, cultural, interactional, and emotional factors that affect all human activity. "The medium is the message!" As will become evident in the next chapter, the implicit values and underlying assumptions embedded in action research provide a set of guiding principles that can facilitate a democratic, participatory, liberating, and life-enhancing approach to research.

COMMUNITY-BASED ACTION RESEARCH: PARTICIPATORY APPROACHES TO INQUIRY

Historically, community-based action research is related to models of action research that sought to apply the tools of anthropology and other disciplines to the practical resolution of social problems (e.g., Goodenough, 1963; Lewin, 1946). Action research ultimately suffered a decline in favor because of its association with radical political activism in the 1960s. In the past two decades it has reemerged in response to both pragmatic and philosophical pressures and is now more broadly understood as "disciplined inquiry (research) which seeks focused efforts to improve the quality of people's organizational, community and family lives" (Calhoun, 1993, p. 62). Action research was also allied to the emergence of practitioner research (e.g., Anderson, Herr, & Nihlen, 1994), new paradigm research (Reason, 1988), and teacher research (e.g., Kincheloe, 1991). The approach to research suggested by community-based action research is implied within the methodological frameworks of fourth-generation evaluation (Guba & Lincoln, 1989). Their dialogic, hermeneutic (meaning-making) approach to evaluation implies a more democratic, empowering, and humanizing approach to inquiry, which is the ideological basis for community-based action research. Since that time there has been a proliferation of texts that speak to a wide range of audiences in professional, organizational, and community contexts (see following discussion), providing a rich resource that, over time, may transform the notion of research.

As an evolving approach to inquiry, action research envisages a collaborative approach to investigation that seeks to engage "subjects" as equal and full participants in the research process. A fundamental premise of community-based action research is that it commences with an interest in the problems of a group, a community, or an organization. Its purpose is to assist people in extending their understanding of their situation and thus in resolving problems that confront them. Put another way, community-based action research provides a model for enacting local, action-oriented approaches to inquiry, applying small-scale theorizing to specific problems in specific situations (Denzin & Lincoln, 1994).

Action research is always enacted in accordance with an explicit set of social values. In modern, democratic social contexts, it is seen as a process of inquiry that has the following characteristics:

- It is *democratic,* enabling the participation of all people.
- It is *equitable,* acknowledging people's equality of worth.
- It is *liberating,* providing freedom from oppressive, debilitating conditions.
- It is life *enhancing,* enabling the expression of people's full human potential.

Community-based action research works on the assumption, therefore, that all stakeholders—those whose lives are affected by the problem under study—should be engaged in the processes of investigation. Stakeholders participate in a process of rigorous inquiry, acquiring information (collecting data) and reflecting on that information (analyzing) to transform their understanding about the nature of the problem under investigation (theorizing). This new set of understandings is then applied to plans for resolution of the problem (action), which, in turn, provides the context for testing hypotheses derived from group theorizing (evaluation).

Collaborative exploration helps practitioners, agency workers, client groups, and other stakeholding parties to develop increasingly sophisticated understandings of the problems and issues that confront them. As they rigorously explore and reflect on their situation together, they can repudiate social myths, misconceptions, and misrepresentations and formulate more constructive analyses of their situation. By sharing their diverse knowledge and experience—expert, professional, and lay—stakeholders can create solutions to their problems and, in the process, improve the quality of their community life.

The role of the research facilitator, in this context, becomes more facilitative and less directive. Knowledge acquisition/production proceeds as a collective process, engaging people who have previously been the "subjects" of research in the process of defining and redefining the corpus of understanding on which their community or organizational life is based. As they collectively investigate their own situation, stakeholders build a consensual vision of their lifeworld. Community-based action research results not only in a collective vision but also in a sense of community. It operates at the intellectual level as well as at social, cultural, political, and emotional levels.

INQUIRY IN USE

A colleague approached me after listening to my report on one of the action research projects in which I had been involved. "You know," she said, "the difference with your work is that you expect something to actually happen as a result of your research activities."

My colleague's statement characterizes, for me, one of the significant differences between action research and traditional research. Traditional research projects are complete when a report has been written and presented to the contracting agency or published in an academic journal. Community-based action research can have these purely academic outcomes and may provide the basis for rich and profound theorizing and basic knowledge production, but its primary purpose is as a practical tool for solving problems experienced by people in their professional, community, or private lives. If an action research project does not *make a difference*, in a specific way, for practitioners and/or their clients, then it has failed to achieve its objective. The analogue of hypothesis testing in action research is some form of change or development that is tested by its ability to enhance the lives of the people with whom it is engaged.

Community-based action research has been employed successfully in schools, hospitals, health clinics, community agencies, government departments, rural communities, urban and suburban organizations, churches, youth clubs, ethnic groups, extension services, and many other settings. It has been used with factory workers, agency staff, school students, youth groups, young mothers, senior citizens, poor people, persons who are unemployed, community groups, people with particular forms of disability or illness, and so on. Action research has been successfully facilitated by welfare workers, social workers, community workers, teachers, nurses, doctors, managers and administrators, urban and community planners, and agency workers in a wide range of social contexts.

The ability of ordinary people to engage in complex organizational work usually deemed the province of professionals has been demonstrated many times. One of the most striking examples I have seen was a community school set up by an Aboriginal group in a remote region of Australia. Weary of sending their young children 150 miles away to the nearest town for schooling, members of the community asked a young teacher to assist them in developing their own school. Untrained for this specialized task, she nevertheless worked with members of the community for some months to build the school from the ground up. Together, they formulated the curriculum and timetable, acquired teaching/learning materials and equipment, secured funding, learned how to satisfy legal and bureaucratic requirements, and built a large, grass-covered hut for a school building. When this small school commenced operation, all classes were taught in one room, with community members helping to teach academic subjects, art, music, and language. The cultural style of the classroom was distinctively Aboriginal, with children happily and busily interacting in small groups, their work supervised by community members and the non-Aboriginal teacher. It was the most successful Aboriginal school I have seen in regard to the enthusiasm and engagement of the children and the sense of energy and excitement that typified the school's operation. Most striking, however, was the sense that community members considered it to be their school and the degree to which they continued, through an extended period, to invest their meager financial resources and considerable time and energy in its operation.

Since I saw that school in operation I have come across many other contexts, including those in the United States, where teachers collaborated with their students, parent groups, and/or colleagues to make deep-seated changes in their schools and classrooms. I have seen striking work in an urban classroom, a successful school comprised largely of Hispanic high school dropouts, and transformative processes in an elementary school in a poor Hispanic neighborhood. What I initially saw in outback Australia seems to have applications in highly diverse contexts.

The following chapters present a set of routines intended to provide guidance for practitioners who wish to engage in action research. Readers may use this approach to inquiry to do several things:

- Enhance everyday work practices by
  - *Reviewing* goals and procedures (What things are we doing? How are we doing them?)
  - *Evaluating* effectiveness (To what extent are we achieving our objectives? How effective is our work?)
  - *Planning* activities and strategies (What needs to be done? How do we get it done?)
- Resolve specific problems and crises by
  - *Defining* the problem
  - *Exploring* its context
  - *Analyzing* its component parts
  - *Developing* strategies for its resolution

- Develop special projects and programs by
  - Planning
  - Implementing
  - Evaluating

Health professionals, for example, may wish to investigate and remediate poor health conditions or practices with a particular community group (e.g., smoking, drinking, low birth weight, or inappropriate medication) and to develop appropriate remedial strategies. School teachers may investigate strategies for dealing with low student achievement levels, poor attendance, student disinterest, or disruptive behavior. School principals may formulate programs for increasing community participation in their schools. Welfare workers may seek to act on the prevalence of child abuse or neglect among client groups. Community workers may wish to develop programs and projects to deal with the problems of neighborhood youth. All will benefit from the use of procedures that enable them to explore systematically the conditions that operate in their specific contexts and that help them develop practical plans for dealing effectively with the problems that confront them.

Some of the areas in which the application of community-based action research may be fruitful, therefore, include the following:

- Education
  - School improvement plans
  - Curriculum development
  - Evaluation
  - Classroom instruction
  - Class projects
  - Special programs
  - Parent participation
  - Site-based management

- Health care
  - Care plans
  - Case management
  - Health promotion
  - Community health projects
  - Community health services

- Social work
  - Child protection
  - Assessment and evaluation
  - Youth programs
  - Parenting programs
  - Community development

- Organizational development
  - Review
  - Planning
  - Change process
  - Project and program development
  - Training programs
  - Cross-cultural programs
  - Human resource development

- Planning and architecture
  - Urban planning projects
  - Community planning projects
  - Housing development projects
  - Housing needs surveys
  - Youth housing needs

THE LITERATURE ON ACTION RESEARCH

The literature on action research is now extensive, indicating the extent to which it is applied to a wide range of professional and organizational contexts. The literature not only provides insight into different approaches to action research but also presents many examples and case studies of the way people have applied it in different contexts.

A variety of journals supply a rich body of resources for those intent on exploring the diversity of approaches to action research and the different contexts in which it is applied. Major sources include the *Action Research Journal*, *ALAR—The Action Learning/Action Research Journal*, the *International Journal of Action Research*, and *Educational Action Research*. Online journals include *Action Research International*, *AR Expeditions*, and, in education, the *Ontario Action Researcher*.

A central resource for action research is the *Handbook of Action Research* (Reason & Bradbury, 2007), which explores the philosophical and theoretical underpinnings of action research, as well as provides case studies from a number of fields. Other general texts include those by McNiff and Whitehead (2006), Whitehead and McNiff (2006), and Fals-Borda and Rahman (1991), and an action research planner by Kemmis and McTaggart (1999). McTaggart (1997) describes the application of action research in an international context, while Heron (1997) provides a detailed and practical guide to cooperative inquiry, one of the many variants of action research. Tyler (2006) and Reeb (2007) present community-focused approaches to action research, and Greenwood and Levin (2006) and Schmuck (2006) present action research as a tool for social change.

Education is perhaps the most prolific source of action research resources. Holly, Arhar, and Kasten (2004) show how action research can be used for curriculum purposes and incorporate case studies in classroom contexts. Mills (2006) positions action research as a fundamental component of teaching, alongside curriculum development, assessment, and classroom management, and Stringer (2004) applies action research more broadly to classroom teaching, student research, and work with families and communities. Other education sources include Noffke and Stevenson (1995); McNiff, Lomax, and Whitehead (1996); Hendricks (2006); McNiff and Whitehead (2006); and Sagor (2004).

In the health sciences, Stringer and Genat (2004) present detailed procedures and case studies of action research in health contexts. Hart and Bond (1995), Morton-Cooper (2000), and Koch and Kralich (2006) provide practical guidance for the use of action research in health care practice. Winter and Munn-Giddings (2001) present a synthesis of the theories and principles guiding action research, and Benner (1994) provides a phenomenological perspective on research health issues that is consonant with action research.

In other fields, Stringer and Dwyer (2005) present action research processes particularly relevant to social workers and other human service professionals, and Coughlan and Brannick (2004), McMurray and Pace (2006), and Whitehead and McNiff (2006) focus on the conduct of action research in organizational settings.

These citations, however, provide only a sample of the literature on action research. It continues to proliferate and is complemented by an increasingly large array of Web-based resources (see Appendix B).

Reflection and Practice

The questions at the end of each chapter provide opportunities for readers to reflect on the issues presented to assist them in "making sense" of the content of the chapter. Individual readers may relate the issues to their own experience to ensure that they understand the concepts, procedures, and techniques that have been described. The process of clarifying and understanding will be enhanced if they engage in conversations around these issues with friends, colleagues, or classmates.

This section also includes activities that provide opportunities for readers to practice some of the techniques and procedures presented. Practice the activities suggested, reflect on what you have done, and have others observe you, if possible. The process of reflecting and obtaining feedback on your actions will be informative.

If you cannot perform all the activities, choose those with which you have least experience or understanding, or which appear to be most useful for your purposes.

Think about the following questions:

- What are the main features of action research?
- Can you describe how action research relates to other forms of research? How is it similar? How is it different?
- Think about a place you work or where you engage in significant activities (clubs, church, friendship groups, etc.). What are some of the issues and problems you or others experience in this context?
- Do you have any ideas about how to resolve any of the issues or problems?
- Write down the place, the issues/problems, and the way you think you might resolve them. Keep these notes for later activities.
- Discuss these issues with a group of colleagues or classmates.

NOTE

Throughout this book, I use my own experiences to illustrate points made in the text. These sections, *biographical bulletins,* are set off from the text with a different typeface.

BOX 1.2

Action Research in Professional and Public Life

Purpose

This chapter introduces the reader to action research as an approach to inquiry.

Content

The chapter presents:

The purpose of action research and where and how it is applied

How action research acts as a systematic process of inquiry

A basic action research routine

An understanding of the participatory nature of action research

The ways that action research is used by diverse groups of professional practitioners and community groups

An introduction to the literature on action research

TWO

THEORY AND PRINCIPLES OF ACTION RESEARCH

INTRODUCING THE THEORETICAL FOUNDATIONS OF ACTION RESEARCH

Chapter 1 alluded to the underlying philosophical stance of action research, revealing the emergence of the tradition and its relationship with other research paradigms. Fundamentally, action research is grounded in a qualitative research paradigm whose purpose is to gain greater clarity and understanding of a question, problem, or issue. Unlike quantitative research (sometimes referred to as experimental or positivistic research) that is based on the precise definition, measurement, and analysis of the relationship between a carefully defined set of variables, action research commences with a question, problem, or issue that is rather broadly defined. Investigations therefore seek initially to clarify the issue investigated and to reveal the way participants describe their actual experience of that issue—how things happen and how it affects them.[1]

In these circumstances, action research is necessarily based on localized studies that focus on the need to understand *how* things are happening, rather than merely on *what* is happening, and to understand the ways that stakeholders—the different people concerned with the issue—perceive, interpret, and respond to events related to the issue investigated. This does not mean that quantitative information is necessarily excluded from a study, because it often provides significant information that is part of the body of knowledge that needs to be incorporated into the study. This information can be included in the processes of meaning making that are essential to action research, but it does not form the central core of the processes of investigation.

This research stance acknowledges the limitations of the knowledge and understanding of the "expert" researcher and takes account of the experience and understanding of those who are centrally involved in the issue explored—the stakeholders. By so doing, researchers take into account a central reality of social life—that all social events are subject to ongoing construction and negotiation. By incorporating the perspectives and responses of key stakeholders as an integral part of the research process, a collaborative analysis of the situation provides the basis for deep-seated understandings that lead to effective remedial action.

Formally, then, action research, in its most effective forms, is phenomenological (focusing on people's actual lived experience/reality), interpretive (focusing on their interpretation of acts and activities), and hermeneutic (incorporating the meaning people make of events in their lives). It provides the means by which stakeholders—those centrally affected by the issue investigated—explore their experience, gain greater clarity and understanding of events and activities, and use those extended understandings to construct effective solutions to the problem(s) on which the study was focused.

These processes do not, however, occur in a socially neutral setting, but are subject to deeply seated social and cultural forces that are taken into account through the participatory processes of investigation of a community-based action research. A more extensive exposition of these forces is provided in the final chapter of this text.

THE CULTURAL STYLE OF ACTION RESEARCH: CAPACITY-BUILDING PROCESSES

Community-based action research seeks to change the social and personal dynamics of the research situation so that the research process enhances the lives of all those who participate. It is a collaborative approach to inquiry that seeks to build positive working relationships and productive communicative styles. Its intent is to provide a climate that enables disparate groups of people to work harmoniously and productively to achieve a set of goals. It is fundamentally a consensual approach to inquiry and works from the assumption that cooperation and consensus should be the primary orientation of research activity. It links groups that potentially are in conflict so that they may attain viable, sustainable, and effective solutions to problems that affect their work or community lives through dialogue and negotiation.

The payoffs for this approach to research are potentially enormous. Not only do research participants acquire the individual capacity to engage in systematic research that they can apply to other issues in other contexts, but they also build a supportive network of collaborative relationships that provides them with an ongoing resource. Solutions that emerge from the research process therefore become much more sustainable, enabling people to maintain the momentum of their activity over extended periods of time. Links established in one project may provide access to information and support that build the power of the people in many different ways.

There are many examples of the way this can operate. I've seen highly effective classrooms where teachers organized students into collaborative work groups to investigate ways to clearly define their learning goals and formulate strategies to accomplish them. I've seen community nurses engage people with chronic health conditions to assist them in establishing ways to live more comfortably and deal effectively with issues confronting them in their day-to-day lives. I've seen youth workers accomplish wonderfully effective programs for marginalized youth that have transformed the communities in which they lived. At the heart of all these activities has been a process of discovery involving the people themselves—clients, students, local youth groups, and so on. In each case the people acquired the capacity to become self-directed and self-sufficient, acquiring a supportive group of peers who could assist them and support them as they engaged the tasks before them.

The effect on the people themselves often has been quite dramatic. At the completion of one action research project, I asked the women in the participating neighborhood group about their experience. One burst out excitedly, "It was such an empowering experience!" As they explored this comment further, it was clear that the women had really appreciated the facts that people had listened seriously to their viewpoints, that they had learned so much, and that they had been actively involved in the research. Originally this project was to have been carried out by research consultants, but with assistance from a local university professor, the members of the neighborhood group had engaged in a "survey" of parent and teacher perspectives on a school issue. The result not only provided the basis for ongoing developments within the school but provided members of the neighborhood group with the capacity and desire to apply their newfound knowledge to a project in a local high school. This was clearly a case where participants had built their capacity to engage in research, as well as increasing the capability of the school to engage in much-needed changes to procedures for communicating with parents.

Building people's capacity to attain significant outcomes that are sustainable over time requires everyone concerned to take into account the impact of activities on all of the issues affecting their lives. It is not sufficient to try to "get the job done" if underlying conditions inhibit or prevent the work in which people are engaged. Too often we have come to accept the impersonal, mechanistic, and allegedly objective procedures common to many contexts as a necessary evil. We endure hierarchical and authoritarian modes of organization and control despite the sense of frustration, powerlessness, and stress frequently felt by practitioners, client groups, students, or others affected by underlying issues within the situation.

It is not difficult to understand why people often suffer deep-seated feelings of frustration. They feel that the centrally controlled programs and services they must institute cannot take into account the multitude of factors that impinge on people's lives. Professional practitioners often feel compelled to implement programs and services according to formally sanctioned practices and procedures, despite their ineffectiveness in achieving the goals they were designed to accomplish. Welfare services often maintain people in states of dependency, passive and irrelevant learning processes are common in many school systems, and medical practices frequently maintain high levels of dependency on prescription drugs. All are symptomatic of underlying disorders embedded in generally accepted practices in our social institutions, agencies, and organizations.

In my professional life, I have often seen programs that isolate people from their families or communities. I have seen services that demean the recipients and organizations and agencies that operate according to rules and regulations that are shamelessly insensitive to the cultures of their clients. I have seen young children isolated from their families for months, sometimes years at a time, in order to be given a "good education." I have seen police fail to act on violence against women because the women were drunk. I have seen millions of dollars wasted on training programs that were purportedly designed to serve community needs but that failed to reach the people for whom they were formulated. I have seen health clinics that were incapable of serving rural community needs because they operated according to practices common in city hospitals. In one community, I was shown the boys' and girls' hostels for high school children, isolated from each other by the length of the town, with the girls' hostel protected by a barbed-wire fence. The administrator who showed me these institutions was proud that "we haven't had an illegitimate pregnancy in years" and seemed unaware of the potential for enormous damage to family and community life inherent in the situation.

The list goes on and on, reflecting the failure of centrally controlled social, educational, health, welfare, and community services to adapt and adjust their operations to the social, cultural, and political realities of the specific locations where they operated. I grieve for the people who have been damaged in the process, including those workers who have become hardened to the plight of the client groups they serve.

I have seen other situations, however, in which administrators, professional practitioners, and workers engaged the energy and potential of the people they served to develop highly effective programs and services. I have seen women's groups that provided for significant needs within their communities, police initiatives that greatly enhanced the peacekeeping mission of the department, and health programs that greatly reduced the incidence of trachoma. I have also seen education and training programs traditionally shunned by marginalized groups become so successful that they were unable to accommodate the numbers of people requesting entry. I have also applauded community youth programs that were able to unite hostile community factions to diminish the problems of young people in their town. I rejoice in them. They have in common a developmental process that maximized the participation of the people they served.

I have written elsewhere of the success of an independent school started by the community in which it operated. It stands in stark contrast to another school I visited. The principal, hired by an outside agency, proudly related the story of the new high school he had set up. With little assistance or support he had organized the renovation of the school building, bought the furniture and equipment, designed the curriculum, and hired the teachers. "I have only one major problem," he confided. "I can't get the parents to show any interest."

As practitioners develop programs and services, therefore, or seek to solve problems that threaten the efficacy of services for which they are responsible, they need to take into account the impacts of those developments and solutions on the lives of the people they serve. Tony Kelly and Russell Gluck (1979) proposed that programs be evaluated not only according to their technical or functional worth but also according to their impact on people's social and emotional lives. Their evaluative criteria investigate the effects of our research activities on

- *Pride:* people's feelings of self-worth
- *Dignity:* people's feelings of autonomy, independence, and competence
- *Identity:* people's affirmation of social identities (woman, worker, Hispanic, etc.)
- *Control:* people's feelings of control over resources, decisions, actions, events, and activities
- *Responsibility:* people's ability to be accountable for their own actions

- *Unity:* the solidarity of groups of which people are members
- *Place:* places where people feel at ease
- *Location:* people's attachment to locales to which they have important historical, cultural, or social ties

These concepts echo features described by Egon Guba and Yvonna Lincoln (1989) in their book *Fourth Generation Evaluation.* Their hermeneutic dialectic process, or meaning-making dialogue, requires interactions that respect people's dignity, integrity, and privacy through

- Full participatory involvement
- Political parity of those involved
- Consensual, informed, sophisticated joint construction
- Conceptual parity
- Refusal to treat individuals as subjects or objects of study

These principles are consonant with those presented by Gustavsen (2001), who outlines the conditions required for a "democratic dialogue," and with Habermas's (1984) concept of "communicative action".

The implicit assumption in these ideas is that procedural matters that directly affect the quality of people's lives need to be taken into account, not only for humanitarian or ethical reasons but also for underlying pragmatic purposes. As Guba and Lincoln put it so succinctly, attention to these properties is likely to "unleash energy, stimulate creativity, instill pride, build commitment, prompt the taking of responsibility, and evoke a sense of investment and ownership" (1989, p. 227).

THE ROLE OF THE RESEARCHER

The assumptions delineated earlier dramatically change the role of the person who is traditionally called "the researcher." In action research, the role of the researcher is not that of an expert who *does* research but that of a resource person. He or she becomes a facilitator or consultant who acts as a catalyst to assist stakeholders in defining their problems clearly and to support them as they work toward effective solutions to the issues that concern them. Titles such as *facilitator, associate,* and *consultant* are more appropriate in community-based action research than *director, chief,* or *head,* which are common to more hierarchical operations. The language signals the nature of relationships and the orientation of the research.

A group of community workers characterized their community-based work in this way (Kickett, McCauley, & Stringer, 1986, p. 5):

- You are there as a *catalyst.*
- Your role is not to impose but to *stimulate people* to change. This is done by addressing issues that concern them *now.*
- The essence of the work is *process—the way things are done*—rather than the result achieved.
- The key is to enable people to develop their own analysis of their issues.
- Start where people are, not where someone else thinks they are or ought to be.
- Help people to analyze their situation, consider findings, plan how to keep what they want, and change what they do not like.
- Enable people to examine several courses of action and the probable results or consequences of each option. After a plan has been selected it is the worker's role to *assist in implementing* the plan by raising issues and possible weaknesses and by helping to locate resources.
- The worker is not an advocate for the group for which he or she works.
- The worker does not focus only on solutions to problems but on *human development.* The responsibility for a project's success lies with the people.

This "*bottom-up*" or grassroots orientation uses stakeholding groups as the primary focus of attention and the source of decision making. It is an approach that requires research facilitators to work in close collaboration with stakeholders and to formulate "flat" organizational structures that put decision-making power in stakeholders' hands.

The approach differs from the authoritarian manner that is, unfortunately, all too common in institutional and organizational life. People in positions of power tend to assume that, being organizationally superior, they have superior knowledge. We have all experienced or observed the authoritarian teacher, administrator, director, social worker, or expert who is feared and/or mistrusted by clients and organizational inferiors, the autocrat who demands that work be carried out according to his or her dictates. This type of tyranny represents the extreme end of a spectrum of authority but is, to a greater or lesser extent, embedded in the hierarchical operations of many institutions and organizations. Such authority is based on the premise that managers, administrators, policy-makers, or "representatives," because of their knowledge of the big picture, are best able to make decisions for client groups and constituencies.

In many situations, this expectation runs contrary to the lived experience of the people. "Experts" are usually trained in narrow areas and often cannot understand the intricacies and complexities of people's lives. "Representatives" often speak for only one group in a constituency and reflect perspectives and interests of that group alone. Moreover, many of the procedures used to try to broaden inputs from constituent groups are so flawed as to be of little value. Surveys, for instance, are usually limited in scope and are frequently riddled with the agendas, interests, and perspectives of the people who commission or construct them. Further, consultation processes frequently occur under conditions that inhibit the participation of many constituent groups.

A colleague referred to community consultation techniques used by government agencies in his state as processes of "insultation." People usually were asked to attend meetings in places such as schools, public offices, libraries, or upscale hotels, venues that were either inaccessible to people who lacked private transportation or socially or culturally alien to them. Meetings often were held at times or under conditions that prevented the attendance of women with children, employed people, poor people, and members of minority groups. The processes of consultation tended to be dominated by those who happened to have no work or home responsibilities or who had the personal resources to make themselves available at times suited to the institution or the bureaucracy.

I have been present at many meetings at which "representatives" have been asked to provide information that was intended to be used as the basis for significant changes to community life. In one instance, I attended a community consultation meeting with senior politicians and government bureaucrats who wished to gain input on government policy initiatives. It was held at a time when most people in that town were unavailable, so that two vocal young women with little standing in the community became the focus of the consultative process. The only other participants were retired people who were given little information about the nature of the processes in which they were involved and appeared to be puzzled by the discussions that took place.

The politicians and bureaucrats who flew into town about noon flew out fewer than three hours later, apparently happy with their "community consultation." I have little doubt that any action that resulted from that visit was not well received by people in the community. They almost certainly would have been insulted by the paucity of the consultation process and would have perceived any actions as the imposition of outside authority on their lives.

In many situations, individuals tend to react negatively to au processes. Having been subject to the well-intentioned, but often attentions of people in official positions, they protect themselves i ways available to them. Thus, when outside authority is imposed on their lives, they often respond with

Aggression, directed at those who are perceived as controlling their lives

Apathy, which sucks away their vitality and leaves them with feelings of hopelessness or helplessness

Avoidance, which isolates them from the source of authoritarian control

When working with people, we as practitioners need to create the conditions that will mobilize their energy, engage their enthusiasm, and generate activity that can be productively applied to the resolution of issues and problems that concern them.

WORKING PRINCIPLES

Within business and industry and the professions it is now widely recognized that participatory approaches to treatment, teaching, and management are much more effective ways of accomplishing productive and effective outcomes. Participatory processes require a quite different approach to work than that commonly accepted in the past. As indicated earlier, we require a set of practices that take into account the human and social dimensions of the context. No longer can we act as though we are dealing with an insensate machine; now we must acknowledge and take account of people's active, creative, willful, and potentially fractious response to any situation. We need to acquire practices, processes, and skills that enable us to work effectively in this more dynamic situation. The principles outlined in the following paragraphs, therefore, signal some of the issues and factors that need to be taken into account in order to accomplish truly effective outcomes to the sometimes complex and deep-seated problems that affect people's lives.

Action research seeks to develop and maintain social and personal interactions that are nonexploitative and enhance the social and emotional lives of all people who participate. It is organized and conducted in ways that are conducive to the formation of community—the "common unity" of all participants—and that strengthen the democratic, equitable, liberating, and life-enhancing qualities of social life. The principles delineated in the following sections—relationships, communication, participation, and inclusion—can help practitioners to formulate activities that are sensitive to the key elements of this mode of research.

Relationships

The type, nature, and quality of relationships in any social setting will have direct impacts on the quality of people's experience and, through that, the quality of outcomes of any human enterprise. Action research has a primary interest, therefore, in establishing and maintaining positive working relationships (see Box 2.1).

Box 2.1 Relationships in Action Research

Relationships in action research should

- Promote feelings of equality for all people involved
- Maintain harmony
- Avoid conflicts, where possible
- Resolve conflicts that arise, openly and dialogically
- Accept people as they are, not as some people think they ought to be
- Encourage personal, cooperative relationships, rather than impersonal, competitive, conflictual, or authoritarian relationships
- Be sensitive to people's feelings

Key concepts: equality, harmony, acceptance, cooperation, sensitivity

---

A new manager was appointed to supervise the work of a group of social workers with whom I was acquainted. Having little experience in the work of these experienced practitioners, and being ambitious, this manager set out to impress her superiors with her efficiency and effectiveness. She embarked on projects that her staff considered inappropriate and put great pressure on them to work in ways that she perceived to be efficient. In the process, she tried to have them act in ways that were contrary to their previously effective work routines and constantly referred to her superior, the director, when they disputed her direction.

Within a short time, work conditions deteriorated dramatically. The social workers struggled to maintain their operation and, in the process, experienced great frustration and stress to the extent that they started to experience both physical and emotional problems. One staff member took a series of extended leaves, another began visits to a psychiatrist, and another transferred to a different section, her position being filled by a series of temporary workers. Eventually, the manager, also under stress, left the agency, and the entire section was disbanded.

This situation is, unfortunately, not an isolated one. Practitioners who have worked in organizational or institutional settings for any length of time will find the scenario all too familiar.

The force of this type of event is to sensitize us to the need to be consciously aware of the nature of relationships in our everyday professional lives. It suggests the need to reject styles of interaction that emphasize status and power and to move to more consensual modes of operation. It implies the need to develop cooperative approaches to work and harmonious relations between and among people and to reject the aggressive, impersonal, and manipulative relations characteristic of many bureaucratic systems. It emphasizes collegial relationships, rather than those based on hierarchy, and leadership roles that help and support people rather than direct and control them.

When we seek to organize any set of activities within an organizational or community setting, we need to examine the type, nature, and quality of relationships among clients, practitioners, administrators, and other stakeholders. At the base of a productive set of relationships is people's ability to feel that their ideas and agendas are acknowledged and that they can make worthwhile contributions to the common enterprise. This, ultimately, is at the core of the processes of a democratic society.

I am reminded of one of the really fine school principals with whom I served. He was, to me, a leader in the fullest sense of the word. Knowledgeable and skillful, he provided me, as a young teacher, with suggestions for ways to improve my teaching that did not imply that I was not already a capable teacher, suggesting or indicating the areas of weakness in my teaching without making me feel put down; he enhanced my feelings of competence and worth by praising my strengths. He was "Dick" to us teachers much of the time, but became "Mr. Filmer" when the occasion warranted our serious attention, or in the more formal moments of ritualized school activities. The words *gentleman* and *scholar* in their best older senses come to my mind: He was a leader of stature and capability who still provides me with the touchstones by which I evaluate my relationships with colleagues, students, and clients. Thank you, Dick Filmer.

Communication

Action research requires all participants to engage in styles and forms of communication that facilitate the development of harmonious relationships and the effective attainment of group or organizational objectives (see Box 2.2). German scholar Jurgen Habermas (1979) suggests that positive change originates from communicative action that provides people with the capacity to work productively with each other. His formulation of the "ideal speech situation" suggests four fundamental conditions that need to be met if communication is to be effective:

- *Understanding:* The receiver can understand what is being communicated.
- *Truth:* The information is accurate and is not a fabrication.
- *Sincerity:* The communicator is sincere in his or her attempts to communicate and has no hidden agendas.
- *Appropriateness:* The manner, style, and form of communication are appropriate to the people, the setting, and the activity.

Institutional and bureaucratic arenas, because of the nature of their organization and operation, provide many examples in which these conditions are not met. Understanding, for instance, is often inhibited by the use of jargon, complex language, or esoteric subject matter. Professional workers sometimes use technical language that clients either cannot understand or cannot relate to their experience. Academics frequently speak in an idiom that mystifies

Box 2.2 Communication in Action Research

In effective communication, one

- Listens attentively to people
- Accepts and acts on what they say
- Can be understood by everyone
- Is truthful and sincere
- Acts in socially and culturally appropriate ways
- Regularly advises others about what is happening

Key concepts: attentiveness, acceptance, understanding, truth, sincerity, appropriateness, openness

practitioners and laypersons alike. In these instances, understanding is limited, and communication is faulty.

Manipulation through the use of distorted information or failure to make covert agendas explicit is so common that it is often accepted as an unfortunate but necessary part of social, organizational, and political life. Damage to communicative action through untruthfulness, however, often leads to more general problems. When people have been tricked or duped, they are frequently unable to continue to work harmoniously with those they feel have cheated them, and the chances of productive and effective work taking place are diminished accordingly.

In many situations, communication is jeopardized when people feel that the manner or style of the communication is inappropriate or that the person involved is not the appropriate person to be involved. When a person from the majority culture speaks for the interests of a minority group in the absence of an appropriate spokesperson, when an administrator makes decisions about a program or service without consulting his or her staff, and when academic experts with little field experience are responsible for professional training, effective communication is difficult to achieve.

I once argued with one of my female colleagues about the need for members of minority groups to speak for themselves in public forums and to be in control of their own affairs. I was unable to make my point clearly until I asked the question, "Would it be right for a man to be head of the National Organization for Women?" with the implication that a man could control the affairs of that organization, represent the interests of women, and present papers on women's needs at conferences. She saw the point immediately.

Apart from these basic conditions of communication, however, the manner, style, and organization of communicative activity will provide many cues and messages that can have significant impacts on people's feelings of well-being and their orientation to activities and agendas. When people feel acknowledged, accepted, and treated with respect, their feelings of worth are enhanced, and the possibility that they will contribute actively to the work of the group is maximized. Communication is the key to the effective operation of any process of inquiry, providing the means to ensure that people are fully informed of events and activities and have all the information they need to accomplish their work together.

Participation

To the extent that people can participate in the process of exploring the nature and context of the problems that concern them, they have the opportunity to develop immediate and deeply relevant understandings of their situation and to be involved actively in the process of dealing with those problems. The task in these circumstances is to provide a climate that gives people the sense that they are in control of their own lives and that supports them as they take systematic action to improve their circumstances (see Box 2.3).

Box 2.3 Participation in Action Research

Participation is most effective when it

- Enables significant levels of active involvement
- Enables people to perform significant tasks
- Provides support for people as they learn to act for themselves
- Encourages plans and activities that people are able to accomplish themselves
- Deals personally with people rather than with their representatives or agents

Key concepts: involvement, performance, support, accomplishment, personalization

Until historically recent times, people were intimately involved in the production of goods and the delivery of services that were part of their day-to-day lives. The farms that surrounded villages provided employment as well as food and clothing, and storekeeper, smith, mason, and preacher were known to everyone who lived locally. Today, the large and centralized social systems that are characteristic of modern societies alienate people from those who provide for their well-being and from the decisions that affect their lives. People are increasingly subject to the faceless dictates of transnational corporations and state and federal agencies for their employment, health, education, and other aspects of community life.

Centralized control has the advantage of concentrating resources to foster large projects and programs—such as health and education systems—that have the capacity to improve or maintain the quality of life of large populations. The downside of centralized control is that those who determine the "texts" of these projects, programs, and services do so in ways that fit their own interests and agendas and the imperatives of their own social and cultural perspectives. A managerial and professional class dominates decision-making processes, often to the detriment of people from lower classes or cultural minorities. The decision makers determine how things will operate, who will benefit, under what circumstances, and according to which criteria, and services are provided by a professional elite that often has little understanding of the circumstances of the people they serve.

In such situations, there is often a tendency to standardize procedures and practices according to managerial or professional cultural imperatives, on the assumption that these fit the general population, notwithstanding that any large population will vary greatly according to the social groups that constitute it. Social identities related to class, ethnicity, race, religion, age, gender, locality, employment, and leisure activities will relate to divergent lifestyles, beliefs, customs, mores, morals, values, skills, knowledge, understanding, and behavioral propensities. When the social and cultural perspectives of a group diverge significantly from those necessary for the group to take advantage of the standardized procedures developed by central policymakers, problems are likely to occur.

Control always has been a significant factor in bureaucratic life, and supervisory roles increasingly are framed in control-oriented managerial terms. Although there is nothing necessarily problematic about framing leadership activity as management, the language of that discipline tends to reflect the ethos of the corporate world in which it was nurtured. There is a tendency to incorporate the values and agendas of the corporate world—with its emphasis on profit, production, and control—into all areas of social, educational, and welfare life, to frame projects according to economic or technical agendas, for instance, while social, cultural, emotional, and spiritual issues fade into obscurity.

Professional practitioners are currently becoming more aware of the limitations of that expertise, however, and there is an increasing tendency to engage clients, patients, consumers, and students in decision-making processes. We have also become more sensitive to the view that an army of experts is unlikely to be able to meet people's needs if the people themselves remain merely passive recipients of services. As practitioners in many fields now realize, unless people come to understand procedures and practices by participating in their development, any program or service is likely to have limited effects on their lives. Patients who fail to maintain appropriate health practices, passive and disinterested students, recalcitrant welfare recipients, disorderly youth, and families in crisis will often not respond to the authoritative dictates of the "experts" whose task it is to "solve" their problems.

A colleague of mine once had the task of presenting training programs on alcohol and drug abuse. Most of the participants were enrolled under court order, as part of their sentences for drug- and/or alcohol-related offenses. The program, which included information about the physical and psychological effects of alcohol and other drugs, was presented to an audience that was, from my colleague's accounts, almost completely unreceptive. "You could tell that they didn't want to be there, and that they wouldn't believe anything I said to them anyway," he commented. "It was a real waste."

I previously had been involved in a workshop given by a senior academic to a community group that had requested a program that would help them better understand the devastating effects of alcohol consumption. The workshop included exploration of a complex three-factor model of drinking behavior that taxed my intellectual capabilities and required considerable concentration on the part of the other participants. As the workshop progressed, they pointed out that the model was inadequate relative to some of the realities of their community life and suggested modifications that would improve it. All participants worked energetically throughout the afternoon, to the extent that the facilitator commented that he was able to cover more ground in that afternoon than he could in three weeks of course work with his postgraduate students. The energy, involvement, and motivation of the participants reflected their orientation to the processes of the workshop. It made sense from their perspective and spoke to issues that concerned them.

Action research seeks to engage people directly in formulating solutions to problems they confront in their community and organizational lives. It is the researcher's task to facilitate and support these activities, rather than to determine their direction. Leadership, in this instance, is defined according to its function of facilitating organizational and operational processes, rather than defining and controlling them. Active participation is the key to feelings of ownership that motivate people to invest their time and energy to help shape the nature and quality of the acts, activities, and behaviors in which they engage.

Inclusion

Action research seeks to enact an approach to inquiry that includes all relevant stakeholders in the process of investigation. It creates contexts that enable diverse groups to negotiate their various agendas in an atmosphere of mutual trust and acceptance and to work toward effective solutions that concern them (see Box 2.4).

Box 2.4 Inclusion in Action Research

Inclusion in action research involves

- Maximization of the involvement of all relevant individuals
- Inclusion of all groups affected
- Inclusion of all relevant issues—social, economic, cultural, political—rather than a focus on narrow administrative or political agendas
- Ensuring cooperation with other groups, agencies, and organizations
- Ensuring that all relevant groups benefit from activities

Key concepts: individuals, groups, issues, cooperation, benefit

A feature of modern life is the concentration of power in the hands of small groups of people. In public life, "representatives," "leaders," or "managers" are given decision-making power over large groups to enable them to control and organize activities. As a consequence, participation in the organizational lives of schools, health agencies, and other community institutions is often characterized by the dominance of middle-class professional groups and, particularly, administrative interests. Management is greatly affected by the needs to play off the agendas of the various client groups and to deal with political machinations that often arise. In these circumstances, the desire for smoothly administered programs limits the extent to which administrators are willing to tolerate the intimate involvement of their diverse and sometimes antagonistic clients. Administrators therefore often focus decision-making processes on an inner circle of their staff or maintain control through truncated decision-making procedures. Committee membership is often carefully controlled to ensure that only restricted groups of people are involved, and meeting agendas are regulated judiciously.

In these circumstances, the voices of the most powerless groups tend to go unheard, their agendas ignored and their needs unmet. Organizational procedures often operate according to administrative priorities and fail to accommodate the social and cultural imperatives that dominate people's lives. Problems proliferate as practitioners struggle to cope with escalating crises that result from the failure of programs and services to cater to client needs. Moreover, these pressures are sometimes exacerbated by political or community demands that "something be done." All too often, superficial solutions provide the semblance of immediate action but in effect can actually exacerbate the situation.

Lawmakers in the United States have recently enacted legislation that increases penalties for criminal offenses, responding to a growing chorus in the country to send young "criminals" to prison and to extend their sentences. In a country with the third-highest rate of incarceration in the world, the likely outcome of these pressures will be to increase, rather than decrease, criminal activity. I remember the words of a friend of mine who, in his earlier years, had "done time" for a number of offenses. His time in prison had, apart from anything else, increased his criminal skills considerably. He had learned a number of ways to break into and start cars, how to break into stores and houses, how to dispose of stolen goods, and how to evade capture. The major outcome of his prison experience was an increase in the extent of his criminal skills and knowledge.

The potential payoff for opening up the processes of organizational life is enormous. Not only does it provide the possibility of increased human resources; it also creates conditions likely to lead to the formation of operational processes that are socially and culturally appropriate for diverse client groups. By including people in decisions about the programs and services that serve them, practitioners extend their knowledge base considerably and mobilize the resources of the community. Including more people in the process may seem to increase the possibilities for complexity and conflict, but it also enables practitioners to broaden their focus from one that seeks the immediate resolution of specific problems to more encompassing perspectives that have the potential to alleviate many interconnected problems.

Reflection and Practice

Select one of the issues you identified in your reflections at the end of Chapter 1. Reflect further on the issue, using the following questions to guide your explorations. Take notes of the main points that come to your mind.

1. Which people are/were involved with this issue? Who are they?
2. Describe the context in which the issue occurs.

3. What part does each of these people play in the action of this issue? What happens? What do they do?
4. How do the people involved in the issue respond to what is happening? What do they do? What do they say? What do you think they're feeling?
5. What is the nature of the relationship between the people involved?
6. What don't you know about what is going on? What would you like to know?
7. How could you find these things out?
8. What might prevent you from finding these things out?
9. Think about a work situation you have experienced. Who was in charge—the "boss"?
10. What did you like and dislike about the way he/she interacted with you and other staff?
11. How did you respond to that person? What did you think, feel, and do?
12. What does this tell you about how to act when you are in a leadership role? How should you act toward those over whom you have authority?

NOTE

1. Kelly and Sewell (1988) differentiate five approaches to community building: service delivery, advocacy (brokering), community action (campaigning), community development, and intentional community.

BOX 2.5

Working Principles of Community-Based Action Research

Relationships in action research should

- Promote feelings of equality for all people involved
- Maintain harmony
- Avoid conflicts, where possible
- Resolve conflicts that arise, openly and dialogically
- Accept people as they are, not as some people think they ought to be
- Encourage personal, cooperative relationships, rather than impersonal, competitive, conflictual, or authoritarian relationships
- Be sensitive to people's feelings

In effective communication, one

- Listens attentively to people
- Accepts and acts on what they say
- Can be understood by everyone
- Is truthful and sincere
- Acts in socially and culturally appropriate ways
- Regularly advises others about what is happening

Participation is most effective when it

- Enables significant levels of active involvement
- Enables people to perform significant tasks
- Provides support for people as they learn to act for themselves
- Encourages plans and activities that people are able to accomplish themselves
- Deals personally with people rather than with their representatives or agents

Inclusion in action research involves

- Maximizing the involvement of all relevant individuals
- Including all groups affected
- Including all relevant issues—social, economic, cultural, political—rather than focusing on narrow administrative or political agendas
- Ensuring cooperation with other groups, agencies, and organizations
- Ensuring that all relevant groups benefit from activities

THREE

SETTING THE STAGE

Planning a Research Process

DESIGNING EFFECTIVE RESEARCH

One of the major purposes of planning activities is to establish a positive climate that engages the energy and enthusiasm of all stakeholders. Programs, projects, and services that fail to capture the interest or commitment of the people they serve are often ineffective, inefficient, or both. Practitioners often assume that they can implement a program once they have gained official permission, only to discover that client groups refuse to participate. I have spoken with practitioners in many fields who have been frustrated by their inability to engage people in activities that appear, from their own perspective, to be highly desirable. Their comments and questions are indicative of their dissatisfaction:

- How can we get parents to participate in classroom activities? We can't even get them to come to the school.
- These young people must help us if we're to develop this facility for them.
- The mothers should bring their babies to the clinic more regularly, but they won't.
- We just can't seem to stop these kids from drinking or taking drugs or engaging in sex.

The language these practitioners use provides clues to the social dynamics involved. When we try to get people to do something, insist that they must or should do something, or try to stop them from engaging in some activity, we are working from an authoritative position that is likely to generate resistance. Such situations often are characterized by processes in which people in positions of authority already have defined the problem and formulated a solution. They fail to grasp that others may interpret the situation and/or the significance of the problem in ways different from their own or may have different agendas in their lives, with other matters having much higher priority. My experience suggests that programs and projects begun on the basis of the decisions and definitions of authority figures have a high probability of failure.

I have seen many housing projects that failed miserably because of inadequate consultation with community members. In one community I visited regularly, the building coordinator was consistently frustrated in his attempts to hire local labor and failed to understand that the housing project had been imposed in that locality with little attempt to include members of the community in the planning phase. The building authority had met with two community leaders and had made determinations on that basis. Other sections of the community were, quite understandably, somewhat affronted by particular aspects of the plan and failed to invest their energies. Problems multiplied in the community from that time on.

This type of process, however, is common in many arenas. In classrooms I have personally discovered the joy and seen others delighted by the increase in effectiveness of their teaching when they engage their students in processes of planning their own learning. In business and industry it is now broadly accepted that operations and outcomes will be far more effective when workers and clients, at all levels, are included in the processes of investigation and planning. I now recognize the limitations of my own perspective and the power of processes that enable me to take advantage of the resources of the people with whom I work.

In Chapter 2, I introduced the three fundamental steps of a basic action research routine. In the first phase—*look*—participants *define and describe* the problem to be investigated and the general context within which it is set. In the second phase—*think*—they *analyze and interpret* the situation to extend their understanding of the nature and context of the problem. In the third phase of the process—*act*—they formulate *solutions* to the problem. This simple routine, however, masks a complex array of influences and activities. Multiple viewpoints and agendas constantly come into play, resulting in a continuous need to modify and adapt emerging plans. The research task becomes a social process in which people extend and reconstruct information emerging from their inquiry (data and analysis) through continuing cycles of exchange, negotiation, realignment, and repair.

SEEKING CONSENSUS: CONSTRUCTING MEANINGFUL RESEARCH

The art and craft of action research include the careful management of research activities so that stakeholders can jointly construct definitions of the situation that are meaningful to them and provide the basis for formulating effective solutions to the research problem. Egon Guba and Yvonna Lincoln (1989) suggest that "the major task of the constructivist investigator is to tease out the constructions that various actors in a setting hold and, so far as possible, to bring them into conjunction—a joining with one another—and with whatever other information can be brought to bear on the issues involved" (p. 142). *Constructions* are created realities that exist as integrated, systematic, sense-making representations and are the stuff of which people's social lives are built. The aim of inquiry is *not* to establish the truth or to describe what really is happening but to reveal the different truths and realities—construction—held by different individuals and groups. Even people who have the same facts or information will interpret them differently according to their experiences, worldviews, and cultural backgrounds.

The task of the community-based action researcher, therefore, is to develop a context in which individuals and groups with divergent perceptions and interpretations can formulate a construction of their situation that makes sense to them all—a joint construction. Guba and Lincoln (1989) designate this a *hermeneutic dialectic process,* because new meanings emerge as divergent views are compared and contrasted. The major purpose of the process is to achieve a higher level synthesis, to reach a consensus where possible, to otherwise expose and clarify the different perspectives, and to use these consensual/divergent views to build an agenda for negotiating actions to be taken. The hermeneutic dialectic process is fundamental to community-based action research, requiring people to work together with purpose and integrity to ensure the effective resolution of their issues and problems.

Because this activity brings people's work lives, and sometimes private lives, into the public arena, it requires a great deal of tact and sensitivity. Construction, consensus, and negotiation need to take place in conditions that recognize the impacts of these activities on participants' pride and dignity and enhance their feelings of unity, control, and responsibility. The routines described in the following section, therefore, reflect a commitment to the working principles discussed in the preceding chapter—relationships, communication, participation, and inclusion. At every stage of their work, research facilitators should ensure that procedures are in harmony with these guidelines, constantly checking that their actions promote and support people's ability to be active agents in the processes of inquiry. The underlying principle here is that human purposes are at least as important as the technical considerations of research.

LOOKING AT THE LAY OF THE LAND: PRELIMINARY ACTIVITY

The processes described in the following paragraphs are designed to furnish a context that will stimulate stakeholders' interest and inspire them to invest their time and energy. Careful planning enables research participants, including the facilitator, to discover the makeup of the setting and establish a presence in that context.

Establishing Contact

In the early stages of a research project, it is important for facilitators to establish contact with all stakeholder groups as quickly as possible. Each group needs to be informed of events and needs to feel that all members can contribute to the research processes. A stranger who is heard speaking to some groups about issues of concern to other groups may arouse suspicion, antagonism, and fear. With that in mind, research facilitators should place a high priority on identifying stakeholders and informing them of the purpose of their activities.

Initial contact should be informative and neutral. A statement such as the following would be appropriate:

> I'm Ernie Stringer. I've been employed by the Watertown Council to work with unemployed single parents. Some people seem to think there is a problem. I'm not sure if anything needs to be done, but I thought that I might talk with people to find out what they think.

This type of introduction provides a general context for a community-wide conversation. After contact has been established, a process of communication should be formulated, for example, "Thanks for talking with me. You've been most informative. I'd really like to talk with you again. Would that be possible?"

Facilitators should establish convenient times and places to meet with people and should, after initial visits, contact people regularly. This way people are more likely to feel that they are included in the process, that their input is significant, and that the project is theirs in some fundamental sense. This condition of *ownership* is an important element of community-based action research.

When research facilitators talk with people, they should also ask about others who should be included in discussions. This will extend the facilitator's knowledge about the membership of each group and of the different stakeholder groups. Networking, in this context, is not merely a political tool to be used to gain advantage but a social tool that ensures that all stakeholders are included in the process.

Sampling: Identifying Stakeholding Groups

In most research studies, one of the first tasks is to decide which people will be included. In quantitative studies, random selection of participants is the most common method for determining who will take part, but qualitative and action research studies require a different process, often called purposeful sampling, that consciously selects people on the basis of a particular set of attributes. In action research, that major attribute is the extent to which a group or individual is affected by or has an effect on the problem or issue of interest.

Any problem or issue is likely to be affected by a wide range of people. A teacher's work, for instance, is likely to be influenced directly by students, colleagues, administrators, support staff, and parents, but teachers can also be affected by others who come into occasional contact with their classrooms—government officials, police, social workers, community leaders, and church leaders. Health professionals may likewise find their patients influenced by nurses, paraprofessionals, doctors, support staff, patients, and administrators. Social workers and agency staff also find their work with clients affected by others who work in the agency or who provide support services.

Researchers, therefore, need to ensure that all stakeholders—people whose lives are affected—participate in defining and exploring the problem or service under investigation. Although it is not possible for all people to be thus engaged, it is imperative that all stakeholder groups feel that someone is speaking for their interests and is in a position to inform them of what is going on.

Research facilitators should conduct a social analysis of the setting to ensure that all relevant groups are included in the research process. Such a social analysis should identify the groups that have a stake in the problem under consideration, so that men and women from all age, social class, ethnic, racial, and religious groups, in all agencies, institutions, and organizations, feel that they have a voice in the proceedings. It is important to cut across these categories to ensure adequate representation—a middle-class migrant person cannot always speak for lower-class migrant people, an older person for youth, or men for women. Members of groups that do not have voices in the proceedings will likely fail to invest themselves in the research processes and may then undermine any resulting activity.

In many settings, I see middle-class professional people making decisions about programs and services that serve the needs of people about whom they have little understanding. The failure of schools to provide for the learning needs of minority populations signals the failure of school systems to engage minority people in the processes of educational planning. In one Australian town that I know, issues related to Aboriginal students are usually referred to a local Aboriginal pastor, although he represents only one of the Aboriginal groups there. The pastor speaks from the perspective of a Christian Aboriginal person and sometimes neglects the interests of other groups who are more closely bound to their traditional spiritual heritage.

This is not an unusual situation. I have observed many instances in which certain people from a minority group are chosen to speak for all others because their middle-class social skills make them acceptable to mainstream institutions, despite their alienation from the people they are asked to represent.

Youth programs often are set up so that well-mannered, clean-living, decent young people are inducted into courses intended to provide them with leadership skills that will enable them to provide appropriate direction to their peers and thus prevent drug problems, teenage pregnancy, and so on. Participants rarely are chosen by their peers, and few are closely related to the groups of young people who are actively taking drugs or engaging in sex. The real leadership in these situations is ignored in favor of the types of youth preferred by adult community leaders.

I participated in a training workshop in which agency workers sought ways of working with the members of a community group to stop them from drinking so much alcohol. The question asked was, "What can *we* do to stop *them* from drinking?" The answer, I thought at the time, was, "Precious little!"

Social mapping processes can help researchers identify all groups and subgroups affected by the issue at hand so that they can fashion a comprehensive picture of the stakeholders in the setting. Charting the social dimensions of a setting can be useful in enabling people to visualize the diversity of groups in any social setting. All groups may not be involved in the research process, but the charting of stakeholders will help research participants identify those people who are primarily concerned with the issues at hand—sometimes known as the *critical reference groups.* Table 3.1, constructed by researchers who wished to identify the different groups in a Texas school district, is similar to such a chart. Another version of the same table identified gender groups, so that each row was split into male and female. For other purposes, it may be necessary to define categories for age-groups. The information may be recorded in tabular or descriptive form, although a map of the context that locates groups physically is sometimes illuminating.

Sampling: Identifying Key People

Researchers sometimes feel disconcerted when people fail to engage in projects that the researchers consider significant. They are often surprised when obstacles appear without warning, when people fail to attend meetings, when tasks are not performed, or when antagonism is directed toward them. In many such instances, they have not only transgressed the boundaries of people's symbolic territory but failed to obtain permission to enter that territory.

An important preliminary task for research facilitators is to determine the formal structure of relevant organizations. They need to identify and communicate with people in positions of influence and authority and gain their permission to work there, when this is organizationally appropriate. At the same time, they also need to locate informal patterns of influence to ensure that all significant people—sometimes called *opinion leaders* or *gatekeepers*—are included in the early stages of the research process.

Table 3.1 Stakeholder Groups

	Anglo			*African American*			*Hispanic*			*Asian*		
	Lower SES	*Middle SES*	*Upper SES*	*Lower SES*	*Middle SES*	*Upper SES*	*Lower SES*	*Middle SES*	*Upper SES*	*Lower SES*	*Middle SES*	*Upper SES*
Students												
Families												
Teachers												
Administrators												
Businesses												
Churches												
Service agencies												

NOTE: SES = Socioeconomic Status.

In one community project I observed, the agency worker took pains to identify all key people. After two months, he became aware that many of the activities he was promoting were failing to make headway, despite positive responses from many of the townspeople. He eventually discovered that one of the farmers in the district who came to town infrequently was a key person in a number of groups in town. He was highly respected and had a significant impact on community affairs. Once the worker contacted this farmer and talked through the issues with him, many of the obstacles simply disappeared.

One technique for establishing or extending the people to be included in the study is to ask participants, a process sometimes called "snowballing." Researchers may ask, "Who else should we talk with about this issue. Who else is affected by it or has an effect on it?" "Who might provide a different perspective on the issue?" This signals the need to ensure that the people included in the study are drawn from groups or individuals having quite different experiences and perspectives.

Establishing a Role

Traditionally, researchers have carried with them the aura and status of the expert/scientist, expecting and being accorded deference. In action research, however, the role and function of the researcher differ considerably from that model. No longer laden with the onerous task of discovering generalizable truths, the researcher takes on a role that carries with it a unique status that, in a very real sense, must be established in the context of each study.

The facilitator, therefore, first must establish a stance that is perceived as legitimate and nonthreatening by all major stakeholding groups. Problems will soon emerge if the researcher is perceived as a stranger prying into people's affairs for little apparent reason or as an authority attempting to impose an agenda. Although the researcher usually will be there under the auspices of some authority, that fact alone is insufficient to engage the attention or the cooperation of all groups in the setting. In many situations, associations with authority may be a marked hindrance, especially if people perceive that the researcher is there to judge, control, or interfere in their affairs.

Researchers, therefore, need to negotiate their role not only with those in positions of authority but with all the relevant stakeholder groups in the setting.

The development of the role of the research facilitator in these circumstances can be conceptualized as having three elements: agenda, stance, and position.

Agenda

Research facilitators can establish their presence in a setting by informing people of their purpose. Initial introductions should include information, in a nondefinitive form, that invites questions and states the agenda in the broadest terms. Initial statements should provide the tacit understanding that the researcher is a resource person whose role is to assist stakeholders, rather than to prescribe their actions. Researcher roles and functions, therefore, become overtly articulated in the process of introduction. The people involved can begin to understand what the researcher is concerned with and the part he or she plays.

As inquiry proceeds, agendas will begin to emerge and become more clearly formulated. At this early stage, however, it is necessary only to establish the focus of activity in the most general terms.

Stance

A researcher's presentation of self should be as neutral and nonthreatening as possible. Body language, speech, dress, and behavior should be purposeful, inquiring, and unpretentious. The all-knowing stance of the expert, the authoritative demeanor of the boss, or the swagger of the achiever is likely to be detrimental to participatory investigation. Many people, especially those in subordinate positions, are likely to respond negatively to any such stance. Researchers should aim to present themselves in ways likely to be perceived as skilled, supportive, resourceful, and approachable. A friendly, purposeful stance is appropriate for most situations.

Position

Position can be thought of as three entities to which any group in a social setting may stake a claim: physical space, social status, and symbolic territory. A family clinic, although ostensibly available to the whole community, for instance, may be claimed by an informal group known as young mothers. A community center may be *owned* by a particular ethnic group. Part of a school ground may be recognized as belonging to a particular clique of girls. These examples show how informal groups, despite no formal legitimacy, have powerful impacts on social life. When researchers conduct their social analyses, it is imperative that they get to know the interacting networks of social groups and identities that cut across the boundaries of the formal organization of a community, institution, or agency. The physical and symbolic territory that such groups occupy will have significant impacts on the processes of investigation.

Networks of social identities are established more formally in such organizations as schools, hospitals, government agencies, social clubs, businesses, and churches. These institutions carefully delineate activities and allocate the people responsible for each territory. A high school English department, for instance, will operate under the leadership of a senior teacher who is accorded higher status than the other English teachers. The department will have responsibility for teaching English and related subjects and may be located in a particular part of the school building. A hospital trauma clinic will engage in medical emergency work, with orderlies, nurses, and doctors having different tasks and statuses—levels of importance—in that context. Churches will likewise allocate a variety of functions, statuses, and locations for worship, child and youth services, finances, and so on to different groups or individuals.

Difficulties often arise when newcomers or people aspiring to change the status quo attempt to invade or change territory that is already occupied. The following comments illustrate situations in which people perceive that territorial boundaries have been crossed inappropriately:

- But Mrs. Jones and Miss Smith always have done the flowers.
- I'm a history teacher. How on earth can I teach English literature?
- Don Jones is part of my caseload. You have no right to interfere with him.
- Staff have never had access to the overall budget.
- The research methods class has never included action research before.

As researchers commence preparatory work, they should artfully position themselves so that they do not threaten the social space of the people with whom they will be working. They should take a neutral position with regard to their status, activities, and physical location, expressing interest but showing no signs that they have designs on any territory.

Research facilitators also cannot afford to be associated too closely with any one of the stakeholding groups in the setting. Members of all groups need to feel that they can talk freely with facilitators, without fear that their comments will be divulged to members of other groups whom, for one reason or another, they do not trust. They also need assurance that their perceptions and agendas will be considered consequential. Researchers, therefore, should endeavor to spend significant periods in places associated with each of the major stakeholder groups—the practitioner's workplace, the administrator's office, and the places where client services are provided. They should make themselves available constantly in these diverse settings so that all stakeholder groups feel that they have equal access. Visibility is important.

Researchers also should meet with people informally—in coffee shops, lunchrooms, sports arenas, bars, community centers, and even private homes, if invited. As they become known and accepted, they may tentatively request permission to attend events, meetings, activities, and so forth (e.g., "Would it be OK if I came along?"). The more freely researchers are able to participate in the ordinary lives of the people with whom they work, the more likely they are to gain the acceptance crucial to the success of community-based action research. By meeting with people in places where those people feel at home, researchers enable them to speak more freely and, in the process, acknowledge the significance of their cultural settings.

I have seen many mistakes made in all these areas—agenda, stance, and position—by otherwise well-intentioned researchers whose presentation of self effectively distanced or alienated them from those with whom they intend to work. I have seen social workers whose fashionable, high-quality clothing was in stark contrast to the cheap and worn clothing of their clients, serving as a constant reminder of the difference in their status. I have observed stylishly coiffed, manicured, and made-up teachers who were unaware of the subtle yet powerful messages their images presented to parents whose incomes and lifestyles were unable to sustain this type of presentation of self on a day-to-day basis. I have encountered researchers and agency workers who always visited the administrator first, disappearing into an office for extended conversations that raised the suspicions of workers and clients. In one instance, I spoke with a school psychologist who drove to her work in the poorest part of the city in an expensive Mercedes-Benz sports coupe. And in the case of the visiting politicians and bureaucrats mentioned in an earlier chapter, their suits and ties were in embarrassing contrast to the stained and worn clothing of their audience.

The agenda, stance, and positioning of research facilitators thus can have considerable impact on the success, or lack of success, of any community-based action research process. These elements of researchers' presentation carry many implicit messages that affect the extent to which other participants feel at home in the research process and are able to develop feelings of ownership. In summary, facilitators should take care to do the following:

- Present themselves as resource persons.
- Be aware of their dress and appearance.
- Establish their purpose in nonthreatening terms.
- Associate with all groups, formal and informal.
- Be visible.
- Be accessible.
- Meet in places where each of the stakeholder groups feels at home.

CONSTRUCTING A PRELIMINARY PICTURE

Initially, research facilitators should develop an understanding of the setting's social dynamics. They need to identify stakeholding groups, key people, the nature of the community, the purposes and organizational structure of relevant institutions and agencies, and the quality of relationships between and among individuals and groups.

Part of this process involves learning the history of the situation with which the researchers are concerned. This will be done in more detail at a later stage of the process, but researchers will need to know what has gone on with regard to key issues prior to their arrival. Past events sometimes leave legacies of deep hurts and antagonisms that severely limit prospects for successful projects unless they are handled judiciously.

A community agency in a country town contracted with the community services unit for which I worked for help with a problem that threatened to close down the agency's operations. The agency had been set up to provide developmental services to a number of community groups in the town and surrounding region. The government had discontinued funding, however, because it was alleged that the agency was no longer providing the services for which the funds had been provided. This was potentially a serious blow to a small community with limited resources.

We asked the agency executive committee and staff to provide their perspectives on the nature of the problem that had led to the current state of affairs. They provided a general picture of events (which was clarified further as different people presented their own parts of the story) that included the various groups and individuals who had been involved and the parts they had played. Participants in this process included the agency staff—development officer, bookkeeper, secretary, and field officer—and members of the executive committee—chair, secretary, and committee members. They also identified key people in the funding body—the regional manager and the field officers—as well as other individuals and groups who supported or were associated with the agency—members of a church group, the town clerk, and workers in a community agency in a nearby town.

We were given permission to contact each of these groups or individuals and proceeded to meet with them. I spoke first with the field officer of the funding body, and then the regional manager, taking care to maintain a friendly, nonthreatening, yet businesslike relationship and to have each of them describe the situation from his or her perspective. From their points of view, the agency had ceased to function effectively, and staff members had become increasingly hostile each time they were approached to account for funds they had expended. I soon established that relations between the staff of the agency and the funding body were extremely antagonistic, making any communication problematic.

Reestablishing a working relationship between agency staff and officers of the funding body was difficult because agency staff members' attitudes seemed to change constantly. Although we discussed the need to establish a working relationship with the funding body and were able to arrange meetings with relevant people, agency workers would then quickly take a negative stance, criticizing the funding agency and decrying the activities of its workers. We eventually discovered that agency workers were meeting regularly with the agency's former bookkeeper, who fired them up each time they met about the deficiencies and character of the funding body officers.

We met with the former bookkeeper and talked through the issues with him, learning more about the history of the situation in the process and asking him to suggest what could be done. By incorporating his suggestions in our preliminary plan and asking his assistance in carrying them out, we were able to defuse an uncomfortable situation that was hindering the project's progress.

One community worker used questions in the following list to guide the building of his own preliminary picture of a particular situation (McCauley, 1985). It was a picture he kept for his own information only, and it differed from the pictures built by the community groups he worked with, but it served to orient him to the setting and to enable him to keep the "big picture" in mind. He was careful not to impose his picture on others or use it as a tool to define the reality of their situation:

- Relationships
  - Who is related?
  - How are they related?
- History
  - When was the community started?
  - Who were the founders?
  - Why was it started?
  - What has been the main economic base?
  - Are the original families still there?
  - What about the history of minority groups?
- People
  - Important people
  - Elected people
  - Church people
  - Government departments
  - Local government
  - Community leaders
  - Community workers
  - Who is involved?
- Groups
  - Geographic groups
  - Social groups
  - Family groups
  - Community groups
  - Racial groups
  - Ethnic groups
  - Who is involved in each?
  - Who are the key people?
  - Which are the influential groups?
- Problems
  - What problems do you see?
  - What problems do others see?
  - What problems do the power brokers see?
  - What problems do local, state, and federal government officers see?
  - What problems do small groups see?

- Resources
  - What social resources are available?
  - What economic resources are available?
  - Who owns them?
  - Are they owned by individuals, groups, or government?
  - What is missing?
  - Why are they missing?

Research facilitators need to formulate their own questions according to the nature and extent of the inquiry in which they will engage. A more limited context, such as a classroom or agency office, may require a less extensive list of questions. The purpose of such questions is to ensure that researchers gain an adequate understanding of the setting in which they will work.

THE ETHICS OF ACTION RESEARCH

Ethical procedures are an important part of all research, and formal research institutions such as universities, agencies, and organizations all have rules and regulations covering the conduct of research. This arises because researchers in the past have engaged in forms of inquiry that have put subjects or participants at risk. This does not necessarily arise because unscrupulous researchers take advantage of people over whom they have control, though this sometimes happens, but because they unwittingly or carelessly reveal information that put their subjects in situations that are harmful.

The conduct of the Milgram experiments (1963), for instance, in which subjects believed they were administering severe electric shocks to other participants so traumatized them that some required extensive therapy. In anthropological studies, researchers have revealed secret/sacred information that created havoc in the communities concerned. There is a long list of instances where researchers have consciously or unwittingly engaged in practices or procedures that are ethically untenable. As researchers we have a "duty of care" in relation to all people we engage in processes of investigation.

In planning a study, therefore, researchers usually need to take specific steps to ensure that participants come to no harm as a result of their participation in the research project. One of the principal tools to ensure this is to clearly inform them of the purpose, aims, use of results, and likely consequences of the study, a process known as *informed consent.* This requires those responsible for the study to provide written information about the aims, purposes, and processes of the study and to gain written acknowledgment of participants' willingness to participate. Such written consent documents usually include statements indicating that

- People have the right to refuse to participate.
- They may withdraw from the study at any time.
- Data related to their participation will be returned to them.
- Any information (data) will be stored safely so that it cannot be viewed by others.
- None of the information that identifies them will be made public or revealed to others without explicit and written consent.

Because of the participatory nature of action research, ethical considerations work in a special way. The same provisions for duty of care apply, and all stakeholders have the same rights to safety and informed consent that apply in other forms of research. In action research, however, there is a particular imperative to ensure that all participants know what is going on, that the processes are inherently transparent to all. Because participants in an action research process have much more control than is usually accorded participants in a study, they are in effect engaging in a mutual agreement about the conduct of a study. Nevertheless, the need for informed written consent is still required for situations where people are at risk because of the sensitive nature of issues involved in the study.

In some cases, where practitioners engage their students, clients, or patients in routines of inquiry that are part of their everyday engagement, formal consent may not be required. This may be the case in situations where teachers involve students in small projects to find better ways of attaining learning objectives, where social workers engage clients in personal explorations of issues in their lives, or where health professionals assist patients in a systematic exploration of a particular health issue. In these cases, the systematic processes of inquiry are part of the legal framework of duty of professional care, and no formal procedures are required to legitimate them.

Agreement to Participate

Goldburn Community Youth Group

Study of Youth Recreational Needs

Research Facilitator: Lyn Evans—Goldburn Youth Services Coordinator

Phone: 259-0892

The Goldburn Community Youth Group wishes to study the recreational needs of young people in the Goldburn area. The information utilized in this project will be collected by interviewing individual group members. Participants will be asked to engage in two interviews of approximately one-hour duration each. The researcher will also have follow-up sessions with each person to check the accuracy of the information. Names will be kept confidential. Tape recordings and field notes always will be stored in a secure place by the researcher and not used for purposes other than the current study. Information that identifies individuals will only be used with their written permission.

If you wish to withdraw from the study at any time you are free to do so, and if you wish, all information you have given will be shredded or returned.

You may contact the researcher at any time via the above phone number.

I, ___________________, have read the information above and any questions I have asked have been answered to my satisfaction. I agree to participate in this activity on the understanding that I may withdraw at any time without prejudice. I agree that the research data generated may be published provided my name is not used and that I am not otherwise identified.

Signed: _______________ (Participant) Date: _______________

Signed: _______________ (Facilitator) Date: _______________

RIGOR (NOT MORTIS): THE RESEARCH IS ALIVE AND WELL

Much research is confounded by the lack of interest of the participants, but the collaborative processes of action research are designed to promote high levels of enthusiasm and active participation. Enthusiasm and interest, however, should not be mistaken for sound research processes. In their enthusiasm, researchers may fail to engage the *systematic and rigorous* processes that are the hallmark of good research. The basis for rigor in traditional experimental research is founded in commonly established routines for establishing the reliability and validity of the research, but action research, being essentially qualitative, uses a different set of criteria. Rigor in action research is based on checks to ensure that the outcomes of research are *trustworthy*—that they do not merely reflect the particular perspectives, biases, or worldview of the researcher and that they are not based solely on superficial or simplistic analyses of the issues investigated.

Checks for *trustworthiness,* therefore, are designed to ensure that researchers have rigorously established the veracity, truthfulness, or validity of the information and analyses that have emerged from the research process. Lincoln and Guba (1985) suggest that trustworthiness can be established through procedures that assess the following attributes of a study:

- Credibility—the plausibility and integrity of the study
- Transferability—the possibility of applying the outcomes of the study to other contexts
- Dependability—research procedures that are clearly defined and open to scrutiny
- Confirmability—evidence that the procedures described actually took place

Credibility

In action research the credibility of research processes is a fundamental issue. Unless participants are able to trust the integrity of the processes, they are unlikely to make the personal commitments that are essential to a well-founded inquiry. The following issues assist them in gaining this feeling of trust:

- *Prolonged engagement:* A brief interview or conversation does not provide sufficient information to enable people to develop the deep-seated understandings that are the necessary outcomes of the research process. Interviews and focus groups should provide all participants with extended opportunities to explore and express their experience of the acts, activities, events, and issues related to the problem investigated.

- *Persistent observation:* Merely being present in a situation does not count as observation. The credibility of research is enhanced when participants consciously observe events, activities, and the context over a period of time. Consciously observing and taking note of events places a premium on taking note of what is actually happening, rather than describing it from memory, or from an interpretation of what people "think" happened.

- *Triangulation:* The credibility of a study is enhanced when multiple sources of information are incorporated. As Stake (2005) suggests, the inclusion of perspectives from diverse sources enables the inquirer to clarify meaning by identifying different ways the phenomena are being perceived. These perspectives may be complemented and challenged by information derived from observations, reports, and a variety of other sources.

- *Member checking:* Participants are given opportunities to review the raw data, analyses, and reports derived from research procedures. This enables them to verify that the research adequately represents their perspectives and experiences. It also provides opportunities for them to clarify and extend information related to their experience.

- *Participant debriefing:* Member checking may also provide people with opportunities to deal with emotions and feelings that might cloud their vision, inhibit their memories, or color their interpretations of events. Debriefing focuses on the feelings and responses of the participant rather than the information participants have provided.

- *Diverse case analysis:* This criterion relates to the principle of inclusion. Researchers enhance the credibility of the study by ensuring that the perspectives of all stakeholding groups are incorporated into the study.

- *Referential adequacy:* Concepts and ideas within the study should clearly be drawn from and reflect the experiences and perspectives of participating stakeholders, rather than be interpreted according to schema emerging from a theoretical or professional body of knowledge. Reports and other communications therefore should be grounded in the terminology and language of the research participants to ensure that it reflects their perspective and can be clearly understood by them.

Transferability

Unlike traditional quantitative or experimental studies that enable the outcomes of research to be generalized to contexts and groups other than those involved in the research, action research outcomes apply only to the particular people and places that were part of the study. That does not mean, however, that nothing in the study is applicable to others. It indicates, however, the need for procedures that carefully explore the possibility that the outcomes of an action research study may be relevant elsewhere. The basis for that lies in the *detailed description of the context(s), activities, and events* that are reported as part of the outcomes of the study. It is possible for people who were not part of the study to make judgments about whether or not the situation is sufficiently similar to their own for the outcomes to be applied. These judgments indicate the degree of *trust* people have that the research outcomes may be *transferred* to their own situation.

Dependability

Dependability focuses on the extent to which people can trust that all measures required of a systematic research process have been followed. An *inquiry audit* provides a detailed description of the procedures that have been followed and provides the basis for judging the extent to which they are dependable.

Confirmability

Researchers must be able to confirm that the procedures described actually took place. An audit trail enables an observer to view the data collected, instruments, field notes, tapes, journals, or other artifacts related to the study. These confirm the veracity of the study, providing another means for ensuring that the research is trustworthy.

SOCIABLE RESEARCH PROCESSES

When we engage in research, we are not merely employing impersonal, technical routines that do not touch people's lives. The nature of any process of inquiry means that we enter cultural settings that are interactional, emotional, historical, and social. People interact at different times and places for different purposes, bringing histories of experience and understanding that orient them to the setting in particular ways. They suffer stress, enjoy a joke, argue, play, work, and socialize in myriad ways that contribute to the life of the community.

The processes of community-based action research, therefore, are enriched by researchers who contribute to the lives of the groups with whom they work. Researchers increase their effectiveness when they immerse themselves in the richness of group life, talking with people about general events and activities, sharing a birthday cake, participating in informal or leisure activities, telling jokes, and so on. Scientific objectivity is not the purpose here. Although it is important to maintain a neutral stance in relation to interpersonal issues, participation entails the need to develop empathic understandings that come about only through close involvement with people. In these circumstances, community-based action research easily becomes an enjoyable human experience.

I've heard such comments as "All he does is sit around drinking coffee or tea with people"—remarks that, especially in the early stages of a project, contain a grain of truth. I am constantly aware of the symbolic significance of sharing food and drink, so I always take people up on the offer of a cup of tea or coffee, a beer at a bar after work, dinner, a party, or a sporting event. I've been invited to church and into people's homes; I've sat in gardens drinking wine, watched basketball games, and attended school functions.

It's not only that I enjoy socializing—the activity is all part of my work, but it allows me to enter people's lives in a "real" way. These social activities break down barriers to communication, help me develop good working relations with others, and give me greater exposure to the setting in a relatively short time. It's surprising how much work can be done in these contexts. People in the course of social talk will often discuss issues related to the project. They become less guarded in such nonthreatening contexts and reveal much that in more formal contexts they might have thought unimportant.

When this happens, I always make a formal visit to the people concerned soon afterward to ensure that they are comfortable with the conversation. I will say, "I had a great time with your friends yesterday. I enjoyed their company." Then, in the course of conversation, I also mention, "I was interested in what you had to say about. . . . I didn't know that." This enables the other person to check that the confidentiality and safety of our conversations still hold in these contexts. The information is confirmed and often extended, and working relationships are strengthened in the process.

CONCLUSION

The principles outlined in the previous chapter apply particularly in the planning stages of action research. More interactive processes that engage facilitator and other participants in carefully articulated processes of inquiry replace the detached, impersonal procedures associated with traditional research. The sociable nature of the research processes should not, however, mask the necessity for ensuring that rigorous and ethical procedures provide the basis for illuminating and effective investigations. As we do so, however, we need to take account of the ways in which the research study affects people's lives and the ways they respond to it.

When we enact action research processes, we are likely to engage approaches to work and community life that are at odds with the general conventions of the institutions and agencies that form the settings for much of our activity. The open dialogue that constitutes a core ingredient of our research processes runs the risk of disturbing a carefully controlled and regulated social environment. We may unconsciously disturb micropolitical alignments and threaten people who have established positions of power and/or influence. In these circumstances, therefore, it is essential to ensure that our initial entry to the setting is as "soft" as possible. We need to understand the complex interactions among people, events, and activities and to comprehend the various ways in which they interpret their situations so that any activity we initiate sits easily in the minds of the people with whom we work.

Our ultimate goal is to provide a context that enables diverse stakeholders to work collaboratively toward solutions to the significant problems that confront them. The preliminary processes described in this chapter set the stage for this activity. They enable research facilitators to gain an overall picture of the situation, to develop positive relationships with people, and to help them think more deeply about their issues and concerns. The relationships and forms of communication that evolve set the stage for the inclusive and participatory processes that are the basis for common unity and productive action.

Reflection and Practice

Find a "cultural scene" to study—a place you might know a little, but in which you are a "stranger." It might be a store, a school, a classroom, a club or association, a church, or any other public place. (If you have access to Spradley and McCurdy's *The Cultural Experience* [1972] you will get clues about the variety of places people can study.) Your task is to visit that "scene" and make contact with people who are there. Later you will engage in a mini-study of that place.

1. Visit the place. Stay there an extended period (around a half hour or more).
2. Look at what is happening. Introduce yourself to some of the "stakeholders" there.
3. Talk with some of them to find out what happens there.
4. Who are the "key people" in this place?
5. Who are the other "stakeholders"?
6. Why are they there? What do they do?
7. Reflect on this experience.
8. How did it feel to be in a place where you were a "stranger"?
9. How did people respond?
10. How much did you learn? Did you gain a deep or superficial understanding of the place?
11. What would you need to do to gain an in-depth understanding of this place?
12. Discuss this experience with colleagues or classmates.

BOX 3.1

Looking at the Lay of the Land

Purpose

Researchers/facilitators familiarize themselves with the context and design research plans in conjunction with stakeholders.

Process

Engage in Preliminary Activity

Establish contact with people in the setting
Identify stakeholding groups—those affected by the issue to be investigated
Identify key people—those in leadership positions or opinion leaders
Negotiate the researcher's role
Establish the position and role of the researcher(s): within which groups, in which places, doing what activities, with whom
Negotiate research agenda(s): which issue(s) will be the focus of inquiry

Construct a Preliminary Picture of the Context

Stakeholders: Groups and key people
Relationships
History of the situation
Problems
Resources

Ethics

Procedures to ensure well-being of participants
- Privacy
- Confidentiality
- Informed consent

Rigor

Check the trustworthiness of procedures
- Credibility
- Transferability
- Dependability
- Confirmability

FOUR

LOOK

Building the Picture

GAINING INSIGHT: GATHERING DATA

The primary objective of the "Look" stage of the process is to gather information that will enable researchers to extend their understanding of the experience and perspective of the various stakeholders—those mainly affected by or having an influence on the issue investigated. The first cycle of an action research process is therefore qualitative in nature, requiring researchers to gather information about participants' experiences and perspectives and to define the problem/issue in terms that "make sense" in their own terms. We seek to understand participant experiences in order to work toward a viable solution in which people will invest their time and energies. Participants, especially primary stakeholders, are therefore consciously engaged in the process of describing the nature of the problem and gathering information.

This differs from traditional hypothesis testing research in two significant ways. First, participants are knowingly engaged in seeking to develop understandings and solutions, objectivity not being the primary aim of the process, as solutions need to make sense to the subjective experience of the participants.[1] Second, researchers do not hypothesize an answer to the research question but seek to understand the nature of related events—how and why things happen the way they do. Gathering data, therefore, has quite a different purpose.

There is a real sense, however, that we seek a sense of objectivity as we gather data, as we need to ensure that information is gained directly from the participants and is not tainted by the perspectives, biases, or experiences of research facilitators. When we frame research questions for interviews or questionnaires, for instance, we need to be very careful that we don't inadvertently incorporate our own views or ideas. Leading questions that implicitly suggest an answer to the problem taint the research process with the perspective of the researcher.

In the "Think" stage of the research, analysis of information emerging from responses to questions will provide insights from which "interventions"—actions to remedy the situation—are formulated. Continuing research cycles enable evaluation, reformulation, and redevelopment of actions, leading to increasingly effective solutions to the problem at the heart of the research project.

I've seen many examples of action research processes that have provided effective outcomes for the key stakeholders. Lisa, a special education teacher, worked with her students to understand why they were so disinterested in reading. By increasing her understanding of their experience of reading she was able to formulate effective strategies that greatly improved her students' interest and engagement.

In another school in the same city, a group of Hispanic parents expressed concern about the quality of their children's schooling. With the support of the principal they engaged in an action research project that focused on improving communication between the school and the students' families. Volunteers from a Hispanic neighborhood group interviewed parents and teachers to understand their experiences of parent–teacher conferences. They were able to identify a range of ways that both parents and teachers could improve the conferences and in the process also discovered other ways to improve communication between the school and the students' families.

This type of exploration is designed not to gather concrete evidence or objective data but to reveal the reality that makes up people's day-to-day experience, bringing their assumptions, views, and beliefs out in the open and making them available for reflection. The process of talking about their experience provides people with rich insights that enable them to achieve greater clarity about events and activities in their lives. As they share experiences they gain greater understanding of the realities of the situation and the potentials of constructing solutions to the problem that take these realities into consideration.

By working collaboratively, participants develop collective visions of their situation that provide the basis for effective action. At its best, this type of activity is liberating, enabling people to master their world as they see it in a different way—a tangible process of enlightenment.

At a community health workshop I attended, participants were asked to describe the types of drinking behavior displayed by various groups in their community. At the end of the workshop, one of the health workers exclaimed excitedly, "I used to see alcohol as a blanket covering my community. Now I see it in a different way. I can see the different ways that people use alcohol and the different outcomes of that use. Now I can see areas where things can be done." An explanatory framework that focused on descriptions of alcohol use, rather than alcoholism as a whole, resulted in a new vision that provided novel possibilities for action.

This anecdote contains an important message: Problems do not exist in isolation but are part of a complex network of events, activities, perceptions, beliefs, values, routines, and rules—a cultural system maintained through the life of the group, organization, or community. As people reveal relevant details of their situation, they see more clearly the ways in which the research problem or focus is linked to features of their organizational, professional, and/or community lives. This disclosure leads people past their taken-for-granted perspectives and promotes more satisfying, sophisticated, and complete descriptions of their situation.

SOURCES OF INFORMATION

The type of information relevant to an investigation will be determined to a large extent by the nature of the issue or problem investigated. The remainder of this chapter identifies the type of data that might be sought and the processes for gathering that information. As participants explore the issue at hand, they may discover the need for a variety of types of information, and data gathering becomes an ongoing process that emerges as the investigation proceeds. This contrasts with procedures of experimental research in which the data to be gathered are precisely defined in the research design.

A social worker I know set out to investigate why young teenagers in her town engaged in so much vandalism. By talking with some of them she discovered that the underlying reason was boredom, and the sources of boredom included schooling that was boring, the lack of sport facilities, and the closure of parks and gardens. Her investigation broadened to include discussions with the local principal and some teachers, local sports people, the local government office responsible for parks and gardens, and the town police. She discovered that sports clubs lacked facilities to accommodate teenagers, that parks and gardens had been closed for economic reasons, and that the schools were locked into a mandated curriculum that was marginally relevant to the needs and interests of the young people in that region. The information she was able to acquire and share with the stakeholders enabled them to take action to increase leisure activity opportunities for teenagers, with a consequent dramatic drop in acts of vandalism.

I have elsewhere (Stringer, 2004) mentioned Lisa, a special education teacher who was concerned about her students' lack of interest in reading. Initially she tried different teaching strategies to try to increase their engagement, but without success. As she talked with her students, however, she discovered a number of issues that affected their reading activities and was able to take appropriate and successful action to remedy the situation by taking these matters into account. She discovered that different students responded to different types of reading activity, the physical arrangement of seating, the social arrangements related to learning activities (some liked solitary reading activities; others liked to work in groups), and the types of reading material she used. She had not been able to "guess" these in her first attempts, but the appropriate data emerged through collaborative work with her students.

The primary data in action research are derived from interviews with primary and key stakeholders. As the focus of investigation and the issues to be incorporated in the emerging investigation become clearer, other sources of data may become relevant. Research participants may also become involved in events and activities that are themselves informative, or they may systematically observe a site or context. As they do so, a variety of other sources may provide information that further clarifies or extends understanding of the issue being investigated. These may include information derived from

- Interviews
- Focus groups
- Participant observation
- Questionnaires
- Documents
- Records and reports
- Surveys
- The research literature

Researchers need to be parsimonious in selecting information that is specifically pertinent to the issue, as a mass of peripherally relevant information may create less rather than greater clarity.

INTERVIEWS: GUIDED REFLECTION

Interviews provide opportunities for participants to describe the situation in their own terms. It is a reflective process that enables the interviewee to explore his or her experience in detail and to reveal the many features of that experience that have an effect on the issue investigated. The interview process not only provides a record of participants' views and perspectives but also symbolically recognizes the legitimacy of their experience. To initiate an interview, research facilitators should

- Identify themselves, their role, and their purposes
- Ask permission to talk with people and to record information
- Check that the time is convenient for an extended discussion
- Negotiate alternative times and places for interviews, if necessary

Interviews should be characterized as informal conversations. For example, the researcher/facilitator might say, "I'd like to talk with you about how things are going for you, Jenny. Could we have a talk sometime? When could we do that?" Allowing participants to designate the time and place maximizes the possibility that they will suggest contexts in which they feel comfortable. One of the key features of successful interviews is the need for participants to feel as if they can say what they are really thinking, or to express what they are really feeling.

As interviews progress, research facilitators may be presented with viewpoints that appear limited, biased, or wrong. Interviewers should, however, avoid discussion or debate with interviewees, since this detracts from the ability of participants to express views of their own experiences and perspectives. Challenges to participant views will occur naturally as differing perspectives are presented in more public arenas. The task at this stage, to adapt the words of a well-known anthropologist, is "to grasp the natives' point of view, to realize their vision of their world" (Malinowski, 1922/1961, p. 25).

Questions

A major problem with the interview process is that questions are easily tainted by the researcher's perceptions, perspectives, interests, and agendas. Professional practitioners are confronted with practical problems and issues that they usually resolve by applying skills and procedures drawn from their professional stock of knowledge. They are confronted with difficult clients who seem unable or unwilling to follow designated procedures, students who seem unable to perform or behave appropriately, or family or community groups that are dysfunctional. In these instances research always starts with a question that takes a general form: "Why are the clients unable to follow their care plans?"; "Why can't these students learn this material?"; "Why do they engage in these inappropriate/disruptive behaviors?"; or "Why are people in this organization/family always arguing?"

Common professional procedures require practitioners to hypothesize an answer to these questions, formulate a potential solution, and then test it out by having clients/students/groups follow procedures implied by that solution. In action research, the practitioner "brackets"—holds in abeyance—his or her professional stock of knowledge and starts by finding ways of having the client/student/group explore the issue in their own terms. Practitioner perspectives will be included in the research process at a later stage, when primary stakeholders have described both the situation and the issue in their own terms.

Questioning procedures are very delicate, because participants are likely to react negatively if there is an implied judgment or criticism embedded in the question. In the first instance, research facilitators will ask very general questions that enable them to understand the way participants experience the context—"How are things going for you, Jenny?" "Tell me about this school (or your class)"—and following questions enable participants to extend their exploration of their own experience and their own perspectives on issues that emerge in the course of the interview. Spradley (1979) provides a useful framework of relatively neutral and nonleading questions that minimize the extent to which participants' perceptions will be governed by language and concepts inadvertently imposed by researchers.

Grand tour questions are sufficiently global to enable participants to describe the situation in their own terms. They take the form, "Tell me about [your work]," which provides focus without giving direction or suggesting types or forms of responses. Grand tour questions include the following:

- *Typical* questions, which enable respondents to talk of the ways events usually occur (e.g., "How does your group usually work?" "Describe a typical day in your office.")
- *Specific* questions, which focus on specific events or phenomena (e.g., "Can you tell me about yesterday's meeting?" "Describe what happened the last time.")

Researchers can further extend accounts by using activities that enable participants to visualize their situation more clearly:

- A *guided tour* question is a request for an actual tour that allows participants to show researchers (and, where possible, other stakeholders) around their offices, schools, classrooms, clinics, centers, or agencies (e.g., "Could you show me around your center [office/classroom/clinic]?"). Throughout the tour, participants may explain details about the people and activities involved in each part of the setting. Researchers may ask questions as they go (e.g., "Tell me more about this part of the clinic [room/office/class]." "Can you tell me more about the social workers [clients/patients/young people] you've mentioned?").
- A *task* question aids in description (e.g., "Could you draw me a map of the place you've described?"). Maps are often instructive and provide opportunities for extensive questioning and description. Participants can also demonstrate particular features of their work or community lives (e.g., "Can you show or tell me how you put a case study together?").

Further information may be acquired through the skillful use of prompts that enable participants to reveal more details of the phenomena they are discussing:

- *Extension* questions (e.g., "Tell me more about . . ." "Is there anything else you can tell me about . . . ?" "What else?")
- *Encouragement* comments or questions (e.g., "Go on." "Yes?" "Uh huh?")
- *Example* questions (e.g., "Can you give me an example of that?")

Once researchers have established a core of information through grand tour questions, they may gain more detailed information by pursuing a *mini-tour* that focuses on concepts already described (e.g., "You previously mentioned [blank]. . . . Tell me more about that." "Can you describe a typical . . . ?").

Combinations of typical, specific, tour, task, extension, encouragement, and example questions provide opportunities for extensive exploration of the setting, events, and activities.

Research facilitators should take a neutral stance throughout these activities and neither affirm nor dispute, verbally or nonverbally, the information that emerges. At the same time, they should remain keenly attentive, recording responses as accurately as possible. It is essential that they capture participants' own terms and concepts for later use in formulating accounts.

Questions should be carefully formulated to ensure that participants are given the maximum opportunity to present events and phenomena in their own terms and to follow agendas of their own choosing. Researchers should be particularly wary of leading questions that derive from their own interpretive schemata and are not directly related to participant agendas.

Field Notes

Participant researchers should carefully record details of interviews, using field notes or tape recordings for this purpose. Taking field notes is a skill that requires some practice, as it is important to record precisely what is said, using the respondent's language, terms, and concepts. Researchers should resist the impulse to summarize what is said, or record it in terms with which they are familiar or comfortable. This requires researchers to write at speed, so that notes may be rather untidy. If possible, use forms of shorthand in note taking (e.g., *t* for *the*; *takg* for *taking*; and so on).

As an interview commences, or when rapport has been established, the interviewer should ask, "Do you mind if I take a few notes as you speak. I'd like to remember what you've said." As the interview progresses, you might also ask, "Could you just hold on for a moment? I'm a bit behind."

Following the interview, the researcher/interviewer should "member check," reading back the notes and asking whether they are an accurate record of what was said. Alternatively, he or she may choose to type up the notes later and have the respondent read them, preferably with the interviewer present, to check for accuracy.

My own experience suggests that most people enjoy being interviewed using these techniques. They appear to like the fact that someone is listening carefully to what they are saying and consider it of sufficient worth to record it.

The "member checking" process is also informative, providing respondents with opportunities to reflect on ideas or events they have presented and to extend or modify their comments. "Yeah, well, I know I said that, but what I really meant was . . ." "Yes, and now I remember what was going on then. We had been . . . "

Tape Recorders

The use of a tape recorder has the advantage of allowing the researcher to record accounts that are both detailed and accurate. The use of tape recorders can have a number of disadvantages, however, and researchers should carefully weigh the merits of this technology. Technical difficulties with equipment may damage rapport with respondents, and people sometimes find it difficult to talk freely in the presence of a recording device, especially when sensitive issues are discussed. A researcher may need to wait until a reasonable degree of rapport has been established before introducing the possibility of using a tape recorder. When using a recorder, the researcher should be prepared to stop the tape to allow respondents to speak "off the record" if they show signs of discomfort.

Steps for using a tape recorder:

- Prior to the interview, check that the tape recorder is in working order and batteries are charged. Have sufficient tape or recording space for the length of the interview(s).
- Ask the interviewee's permission to record the interview.
- Transcribe the tape as soon after the interview as possible.
- Provide the interviewee with a copy of the transcription to check for accuracy.
- Store tapes and transcriptions in a safe place to ensure confidentiality.

FOCUS GROUPS

Focus groups provide another means of acquiring information and might be characterized as a group interview. Participants in a focus group should each have opportunities to describe their experience and present their perspective on the issues discussed. As with interviews, carefully devised questions provide the means to focus the group on the issue at hand and enable them to express their experience and perspective in their own terms, without the constraints of interpretive frameworks derived from researcher perspectives, professional or technical language, or theoretical constructs. Focus questions should follow the same rules and formats as those used for interviews, employing neutral language and maximizing opportunities for participants to express themselves in their own terms.

The following steps provide a basic framework for facilitating focus groups:

Set ground rules. For example:

- Each person should have opportunities to express his or her opinions and perspectives.
- Participants should be respectful and nonjudgmental of each other.

Explain procedures clearly.

- Designate a leader and recorder for each group (or have the group elect a leader).
- Provide and display focus questions.
- Explain recording and reporting procedures.
- Designate time frames for each activity.

Facilitators should

- Ensure each person has an equal chance to talk.
- Ensure discussions relate to the focus question(s).
- Keep track of time for each activity.
- Assist the group in summarizing the perspectives emerging from their discussions and identify key features of their experience and perspective.

Recorders should

- Keep note of what people say, using their own words.
- Record the outcomes of the summarizing process, preferably on a chart.

Plenary sessions—feedback and clarification:

- Gather groups for plenary session, ensuring adequate time for the process.
- Each group should present the outcomes of its discussion, using the charted summary.
- Where one person presents, provide opportunities for individuals within the group to extend or clarify points presented.
- The facilitator should ask questions that clarify or extend the information presented, extending understanding of the group's perspective.
- Ensure that new information emerging in this process is recorded on charts.

Combined analysis: Have participants

- Identify common features across the charts.
- Identify divergent issues or perspectives.
- Rank issues in order of priority, using some form of voting procedure.

Planning: What happens next?

- Make an action plan, starting with an issue of highest priority.
- Designate tasks, persons, timelines, and resources.
- Designate a person to monitor these tasks.
- Designate a time and place to meet to review progress on the action plan.

PARTICIPANT OBSERVATION

Participant observation requires a form of observation that is distinctively different from observational routines common in experimental research or clinical practice, where the things to be observed are specifically defined. Observation in action research is more ethnographic, enabling an observer to build a picture of the lifeworld of those being observed and an understanding of the way they ordinarily go about their everyday activities. It enables teachers to see how students go about the tasks that have been assigned them, social workers to observe how mothers interact with their children, or consultants to watch how people go about the work of an organization.

As researchers observe stakeholding groups, they will have opportunities to gain a clearer picture of the research context by observing the everyday settings in which participants live and work and the ways they go about their activities. By recording their observations as field notes, they acquire a record of important elements of the lifeworld of the participants. Researchers should record these notes during or soon after events have occurred, especially noting

- *Places:* offices, homes, and community contexts; locations of activities and events; physical layouts
- *People:* individuals, types of people, formal positions, and roles
- *Objects:* buildings, furniture, equipment, and materials
- *Acts:* single actions that people take (e.g., reading a report on a client)
- *Activities:* a set of related acts (e.g., formulating a case study)
- *Events:* a set of related activities (e.g., case conference)
- *Purposes:* what people are trying to accomplish
- *Time:* times, frequency, duration, and sequencing of events and activities
- *Feelings:* emotional orientations and responses to people, events, activities, and so on

Observations enable researchers to record important details that become the basis for formulating descriptions from which stakeholding groups produce their accounts. Although field notes are commonly used for observations, videotapes and photographs may also provide a powerful record of events and activities.

As I talked with people in a community resource agency about their work, I also observed the conditions under which they worked, the resources available to them, the interactions among the different types of workers, and the tasks in which they were engaged. Through time, I gained a much clearer understanding of the operation of the agency than I had initially. I interviewed people and made similar observations during my visits to the funding agency with which the resource agency was in conflict and during visits to other relevant agencies and organizations in town. I recorded information about all these settings, noting those features that seemed relevant to the investigation. I returned to some of the settings to record more details as the need for additional information arose. When agency workers complained of lack of equipment, materials, and stationery, for instance, I was able to assist them in performing a rough inventory and, in the process, was able to build a detailed picture of available resources.

I also visited key people in client groups and other agencies to acquire descriptive information about the general context of the town, the conditions in which its client groups lived, and other details of the community context. This information provided the materials from which I produced a general account of the context, which I presented as a preliminary report at the next meeting of the agency committee. I was careful to present an account that represented the different perceptions and interpretations of the situation given to me by participants in the setting. I asked the committee to discuss each section of the report and then made modifications and additions on the basis of their comments.

Participation in research contexts also provides research facilitators with opportunities to engage in interviews and conversations that extend the pool of information available. Appropriate questions enable researchers to describe the situation in the participants' own terms. For instance,

- What are all the (places, acts, events, and the like)?
- Can you describe in detail the (objects, times, goals, and the like)?
- Can you tell me about all the (people, activities, and the like)? (Spradley, 1979, p. 79)

This process also provides opportunities for researchers to check the veracity of their own observations. Phenomena such as purpose and feeling can be inferred only superficially by an observer and need to be checked for accuracy. I might record, for instance, "The principal met with faculty to present a written statement of the new school policy. The faculty appeared rather disgruntled with the new policy but made no comment about it." Here I have noted information related to the purpose of the meeting and the feelings of the faculty. This is potentially relevant information, but I would need to check with the principal and faculty to verify the authenticity of my interpretation. The principal may have other unstated purposes, the presentation of policy merely providing the context for meeting with faculty, and the faculty themselves may have been disgruntled about those other (hidden) agendas.

DOCUMENTS, RECORDS, AND REPORTS

Researchers can obtain a great deal of significant information by reviewing documents and records. Documents and records may include memos, minutes, records, reports, policy statements, procedure statements, plans, evaluation reports, press accounts, public relations materials, information statements, and newsletters. Organizational literature from schools, hospitals, clinics, centers, churches, businesses, and so on may include client records, policies, plans, strategies, evaluations, reports, procedural manuals, curricula, public relations literature, informational literature, and so forth. These types of documents are often prolific, and researchers need to be selective, briefly scanning their contents to ascertain their relevance to the issue under investigation.

Researchers should inspect official directories, where available, and keep records of the documents reviewed. They should record any significant information found in documents and note its source. In some cases, researchers may be able to obtain photocopies of relevant documents.

Researchers should prepare summary statements of information they have acquired and check them for accuracy with relevant stakeholders. In the process, they should ascertain which information may be made public and which must be kept confidential. The intent of the summaries is to provide stakeholders with information that enables each group to review and reflect on its own activities and to share relevant information with other stakeholding groups. This information will provide the key elements from which a jointly constructed account will be formulated.

SURVEYS

Surveys are of limited utility in the first phases of an action research process, because they provide very limited information and are likely to reflect the perspective, interests, and agendas of the researcher(s). In later stages of action research, however, a survey may provide a very useful tool for extending the data collection process to a broader range of participants. It provides the means to check whether information acquired from participants in the first cycles of a process is relevant to other individuals and groups.

To conduct a survey, researchers should

Establish the purpose and focus of the survey. Carefully define

- The issues to be included
- Information to be obtained
- The respondents to be included in the survey sample

Formulate the questions for the questionnaire:

- Construct *one* question for *each* issue or piece of information.
- State the question in clear and unambiguous terms.
- State the question in positive rather than negative terms.
- Do not include jargon or technical terms that may be unfamiliar to respondents.
- Keep questions short and to the point.

Construct response formats:

- Open response, for example, "What factors have the most impact on your ability to do your work?"
- Fixed response
- Dual response
- Rating response

Provide introductory information for respondents:

- Purpose and nature of the survey
- Likely duration of time to complete responses
- Examples of types of response required

Test the questionnaire:

- Select a small pilot sample of respondents.
- Have them complete the questionnaire or respond to interview questions.
- Analyze responses to identify problematic or inadequate questions.
- Modify those questions.

Conduct the survey:

- Administer questionnaire or interview respondents.
- Collect responses.
- Thank people for their participation.

Analyze the data:

- Collate information for each question.
- Compute appropriate statistical measures—sum, mean, standard deviation, and so on.
- Identify significant results.

Report on the outcomes of the survey.

REVIEWING THE LITERATURE: EVIDENCE FROM RESEARCH STUDIES

In the first stage of action research the purpose of investigations is to extend and clarify participant understanding of the issue. Research participants seek to understand how and why events happen as they do. Often, however, people's "common sense"—the taken-for-granted knowledge that has accumulated through their personal history of experience—comes into conflict with other people in the setting. We need to find ways of checking the validity of these differences in perception, and we also need to check whether commonsense interpretations of the situation can be otherwise verified. Sometimes people assume the truth of a statement that is at odds with other credible sources of information. This is particularly true of perceptions that result from distortions arising from media sensationalist reports or the grandstanding of political figures.

In these situations we have available a stable and powerful body of knowledge, established through a long history of systematic investigations, that enables us to check the veracity or validity of statements that are often presented to us as "facts." While the research literature does not provide definitive answers to all issues, it does provide information that has been thoroughly established through rigorous and systematic studies that provide us with much higher degrees of certainty than can be gained from any other source. When matters of fact are at issue, research can often help resolve disputes or provide information that has a more solid grounding than "What I think is . . ." or "It's just common sense!" Research can assist us in finding the extent to which the following types of assertion have any basis in fact:

- Crime is increasing.
- Students are performing poorly in schools.

- Taking "soft" drugs leads to "hard" drug usage.
- Youth engage in sexual activity (or become pregnant) much younger than they did previously.
- This treatment definitely is successful in creating weight loss (or stopping smoking, improving health, etc.).

The outcomes of research, presented as statistical information, can often provide clear evidence to either support or reject the veracity of such assertions. It is important, however, to check the history of any given set of information to ensure that though things may look bad, they may represent an improvement in the situation. It is also important to check a variety of studies, because single studies often provide inadequate evidence upon which to make judgments. Review reports provide a useful source for information.

Many parents at a school at which I worked were incensed when the principal informed them that she would need to introduce some mixed-grade classes. At a meeting with parents she was able to present strong research evidence that students' academic performance did not suffer when they were placed in mixed-grade classes and that there were a number of psychological and social benefits. Most parents present at the meeting were gratified to have this level of information presented to them and were satisfied that their children's education would not suffer. The strength of the information was adequate to calm fears they had expressed early in the meeting.

A recent politically inspired "law and order" campaign in my city, aimed at greatly increasing legal penalties, was based on statements from political and community figures that indicated an increase in crime. Research that had been recently completed by a local university, however, showed clearly that crime, in almost all areas, had decreased markedly in that period.

Access to the research literature is usually found in university libraries, largely in academic or professional journal reports. Though this is still the greatest repository of research reports, the Internet now provides a number of avenues to enable community-based researchers to acquire this type of information through Web searches. A number of powerful search engines—Google, AltaVista, Yahoo, and so on—and Web sites now provide the means to access information derived from a large body of research. Google Academic provides a general resource, while derivatives of the old ERIC system still provide access to reports and reviews of educational research. Where matters of fact are involved, research participants now have the ability to acquire solid information from the huge array of research currently available (see Appendix B).

To conduct a Web search, researchers should

- Identify and describe the issue or topic.
- Identify key terms that characterize the topic.
- Log on to an appropriate search function. (Click on the "search" function that is incorporated into most computer programs.)
- Enter key terms.
- Review relevant Web sites.

GATHERING STATISTICAL INFORMATION—HOW MANY . . . ?

An action research project may make use of statistical information for a variety of purposes. Records of numbers of events, participants, and so on can contribute to a clearer picture of the status of a research project. An action research project might include the following numerical information:

- Occurrences: The number of people, events, behaviors, participants, and so on
    - A record of the number of people engaged in a project, using a service, participating in activities, and so on provides the means of keeping track of events and estimating the degree to which the project is reaching intended audiences.

- Comparisons: Differences in occurrence between groups
    - Interesting information can be obtained by comparing people from different groups—men and women, people from different age-groups, people of different race/ethnicity, and so on.

- Trends or histories: Changes in occurrence over time
    - A single "snapshot" provides information that, by itself, may be misleading. Sometimes significant accomplishment or success is masked by the fact that a low "score" fails to reveal that considerable improvement has been made over time.

- Central tendencies: The average number of people, occurrences of events, and so on
  - Mean scores help us keep track of occurrences of phenomena, especially where large numbers are involved. We can be better informed of the number of people participating in events, the scores on tests, and so on.

- Distributions: The extent to which scores or occurrences are clustered or widely spread
  - Attention to the distribution of occurrences assists us in understanding whether or not there is a wide dispersal of scores in a sample. The distribution may signify the need to treat groups differently or to pay further attention to individuals or groups who appear to be different in some way, either through the lack of participation indicated by the data or through the outcomes of the activities in which they have been engaged.

EXTENDED UNDERSTANDING: DESCRIPTIVE ANALYSIS

The first phases of an action research process are designed to provide well-grounded understandings of the experience and perspective of participants. Further information from observations and so on provides additional data that can complement, clarify, and extend understanding of the events and other phenomena associated with the issue at hand. Preliminary observations and interviews may lead to more extensive processes that enable participants to construct more sophisticated and detailed accounts of their situation, enabling them to see the complex web of interactions and activities within which problematic events are played out.

The following section therefore provides three alternative procedures researchers may use to assist people in extending their understanding of their situations and the issues investigated. These are similar to ethnographic processes used in qualitative research and have been applied successfully in many community, organizational, and group contexts. Although each process is intended to provide a separate alternative approach to the development of a descriptive account, researchers may find in practice that they include any combination of these procedures. Chapter 5 provides a set of procedures for identifying the features and elements that provide material for developing descriptive accounts.

Alternative 1: Working Ethnographically—Collaborative Descriptive Accounts

The processes outlined in the interviewing section in this chapter may be applied to group contexts. Grand tour and other questions may be directed to a single group or to varied groups attending a meeting rather than to individuals. The research facilitator should elicit multiple responses to these questions verbally, using group processes where appropriate, and should record them in summary form on charts. This material can then be used in the formulation of descriptive accounts (see the relevant section on formulating accounts in Chapter 8).

Alternative 2: Six Questions—Why, What, How, Who, Where, When

The first question—why—provides a general orientation to the focus of the investigation, whereas succeeding questions—what, how, who, where, and when—enable participants to identify associated influences. The intent is not to define causes but to understand how the problem is encompassed in the context or setting. *How* and *what* questions are more productive than *why* questions. The former provide opportunities for revealing direct experience, whereas the latter often lead to forms of explanation that are remote from people's experience. Examples of appropriate initiating questions include the following:

- Why are we meeting today? What is the purpose?
- What is/are the problem(s)? What is happening?
- How does it affect our work/lives?
- Who is being affected?
- Where are things happening?
- When are things happening?

Answers should focus on acts, activities, and events related to the problem; participants should not attempt to evaluate or judge individual or group behavior. An initial pass through these questions will lead to further questions that can provide increasingly detailed information. This will include the history of the situation (how it came to be as it is), the people involved in or affected by the problem, interactions and relationships among these people, resources (people, space, time, funds, current use, access, availability), dreams, and aspirations.

Alternative 3: Community Profile

A community profile provides a descriptive "snapshot" of the context in which the investigation is placed. It enables stakeholders to formulate an overview that describes significant features of their context. Smaller projects in restricted sites, such as classrooms, schools, community centers, and government agencies, may require information that focuses only on dimensions of the setting itself—that is, the classroom, school, agency office, and so on. In many cases, however, persistent problems require investigations that extend into the neighborhood, town, city, or region.

There is often a profusion of information about community contexts, and it is essential that the work of preparing a profile remain focused. Following preliminary activity that briefly describes the setting (see Chapter 3), researchers should ask major stakeholders for their views about the types of information pertinent to the investigation. This will help researchers choose judiciously from a potentially vast array of information and minimize the time spent amassing inconsequential information. Research facilitators should assist stakeholders in developing a community profile framework that appears most appropriate to the task at hand. A community profile might include any of the following:

- *Geography:* location, landforms, climate
- *History:* history of the setting, major events, developments, history of the problem(s) under investigation, major laws affecting the site or the problem(s)
- *Government:* impacts and places of local, regional, state, and federal government policies and agencies; boundaries
- *Politics:* parties, organizations, representatives
- *Demographics:* population size; gender, race, ethnic, and age distributions; births and deaths
- *Economics:* sectors (business and industry, etc.), employment, wages and salary levels, general status (prospering, declining)
- *Health:* services, agencies, facilities, special populations (e.g., older persons, children, persons with disabilities)
- *Education:* schools, institutions, services, sectors (e.g., primary, secondary, college), resources, community education
- *Welfare:* services; institutions, agencies, and organizations (government and nongovernment); personnel (social workers, counselors, etc.)

- *Housing:* number, type, and condition of dwellings; forms of accommodation (e.g., low rent, transient, hotel)
- *Transportation:* public and private transportation, accessibility and availability, areas serviced, road conditions, types (road, rail, air)
- *Recreation:* type, number, condition, and accessibility of facilities and services; clubs and organizations; age-groups targeted
- *Religion:* type and number of churches, levels of attendance, activities and services
- *Intergroup relations:* social groups (race, ethnic, religious, or kinship affiliations), coalitions, antagonisms
- *Planning:* regional, city, town, or local plans

Community profiles usually provide demographic information, much of which can be gained from official or documentary sources, but information may also be acquired during preliminary observations and interviews. John Van Willigen (1993, after Vlachos, 1975) suggests a format that embodies cultural information in addition to the types of demographic data described above. The type of information collected necessarily requires extended ethnographic work with people within the setting. Van Willigen's framework incorporates cultural tracts that include the following:

- *Lifestyle:* economic status, communication (including language, proxemics, and expressive media), religious sites and practices, housing (styles and clustering of dwellings, place of kin networks), geographic location, institutional characteristics, health definitions and practices, education, leisure and recreational activities, politics
- *Historical features:* contemporary and historical artifacts and physical representations
- *Worldviews, beliefs, perceptions, and definitions of reality:* cognitive systems (how people think about and organize everyday reality), religious systems (spiritual reality), values systems, belief systems, perceptions of one's own group and others, intercultural perceptions

Once the researcher has acquired the information for the community profile, he or she can organize and present it in a form that people from all stakeholding groups can readily understand. The profile can be made available for their scrutiny as part of the process of formulating accounts (see Chapter 5).

A community profile provides a structured way for participants to determine clearly the range of influences likely to have an impact on the problem under investigation. The information ensures that a broad range of relevant features of the situation are taken into account and paves the way for effective and sustainable projects and programs.

MEETINGS: GROUP PROCESSES FOR COLLABORATIVE INQUIRY

The most successful and productive action research occurs where individual participants have the opportunity to talk extensively about their experiences and perceptions. Interview processes enable people not only to reveal the issues and agendas but also to reflect on the nature of events that concern them. Where individual interviews are not possible, research facilitators may organize meetings that bring people together to explore the issue under investigation. In these circumstances they should use carefully articulated group processes to ensure that each participant has extended opportunities to describe his or her situation and to express his or her issues and concerns.

Preliminary Meetings

When diverse stakeholder groups are brought together, it is often fruitful, particularly where there has been a history of conflict between parties, for researchers to do some preliminary work to ensure harmonious and productive meetings. Holding prior meetings with each of the conflicting parties can enable them to define their own agendas and to clarify the purposes of the larger meeting. In these contexts, researchers should work to formulate statements that are nonjudgmental and nonblaming, yet clearly articulate participants' perceptions and concerns.

When I worked on community projects, I always indicated the need for joint action between stakeholding groups. After I had established contact with individuals and groups and provided them with opportunities to express their viewpoints, I would signal the possibility of linking with others. I would comment that there are a number of other people (or groups) who are concerned with this issue, and it might be useful to meet with them to discuss it.

A common mistake of researchers intent on community-based action is to call a "public meeting" to discuss issues and formulate plans. Often, such meetings are held in places—schools, universities, agency offices—that are alien to many stakeholders. In consequence, many groups are poorly represented, and the organizers become disconcerted by the apparent apathy of the people they wish to engage. School principals, social workers, and health workers, for instance, cannot understand why certain parent groups will not come to meetings called specifically to deal with their children's problems. They frequently comment in these circumstances on the parents' "lack of interest" and fail to perceive how threatening formal or official environments can be for some people.

I have been involved in many situations in which parents, especially those from lower socioeconomic environments, perceive schools and government offices as threatening and judgmental bastions of authority. Meetings held in institutional arenas often reinforce these perceptions, as articulate and self-confident people hold the floor and determine the course of events and the texts—resolutions, statements, records, and so on—that finally emerge. Persons who are less articulate are effectively silenced, their concerns unheard and their agendas unmet.

In some instances, public meetings provide contexts in which individuals or groups in conflict meet for the first time. Without preliminary work, these types of meetings may degenerate into conflict-laden situations that serve only to reinforce antagonisms and exacerbate existing problems. Public meetings, therefore, should be used only after the various stakeholder groups have had the opportunity to meet in safe and comfortable contexts to explore their issues and to clarify their thoughts and perceptions. Where there is no other alternative, researchers should carefully facilitate group processes, as suggested in the following section, to enable all participants to express their views in safety and have their issues recognized.

Organizing Meetings

Meetings don't "just happen"; careful planning and preparation must take place to ensure that participants can work through their issues and attain their objectives without the distractions of poorly articulated activities, inadequate materials and equipment, or conditions that are uncomfortable or irritating.

Some of the critical issues that research facilitators need to take into account when organizing meetings are discussed in the following section.

Participants

Meetings should reflect the participatory intent of community-based action research; it is important, therefore, to ensure that people who can legitimately speak for the interests of each stakeholding group attend. Having one or two persons represent a diverse ethnic group, for instance, may ignore the deep divisions that lie among various families, cliques, and social class groupings within that population. Researchers should review their initial social analysis to confirm that all groups are appropriately represented by individuals who can legitimately take on the role of spokesperson. Meeting conditions—time, place, and transportation—should maximize the opportunities these people have to attend.

Leadership: Facilitating a Productive Meeting

A meeting is best led by a neutral chair or facilitator—a person perceived as having no overriding loyalty to any particular stakeholding group. Researchers can act as facilitators, but it may be appropriate in some circumstances for a respected community member to act in this role. It is important that participants accept the chair or facilitator as a legitimate or appropriate person to direct proceedings. Only in exceptional circumstances, for example, would a meeting investigating women's issues have a male chair.

The task of negotiating diverse perspectives in a research process can become difficult if strong and determined people try to impose their own agendas and perspectives. The chair or facilitator should employ judicious, diplomatic, yet firm processes to ensure that such people do not stifle the diverse agendas and perspectives that are essential components of the process. For that reason, researchers should ensure that the chair or facilitator has the formal authority and the procedural skills to maintain the integrity of the meeting process. A formal statement, written or verbal, that acknowledges the chair or facilitator's authority should be obtained from key stakeholders prior to the commencement of the meeting. This may take the form of a letter or memorandum to participants or may be included in a welcoming speech at the meeting.

Significant differentials of power between the research facilitator and other participants are not conducive to productive meetings. Researchers should ensure that people of appropriate status are engaged to facilitate or lead meetings.

I learned of a situation in which an inexperienced junior professor and a graduate assistant were asked to facilitate a workshop at a conference of senior educational administrators. Within a short time, the administrators had marginalized the professor and graduate assistant and had taken control of proceedings.

Ground Rules and Agenda

Researchers may ask groups in advance to suggest meeting ground rules in order to minimize the possibility of conflict and to provide conditions conducive to productive work. Meeting procedures should be planned carefully to minimize the possibility that they may degenerate into heated debate, accusation, and blaming. Each meeting should begin with the presentation of a broad agenda that includes statements about (a) the purpose of the meeting, (b) the manner in which the meeting will proceed, and (c) the activities in which participants will engage. Time may be allocated for people to comment on the agenda and to suggest alternative procedures. These preliminary activities, however, should not take so much time that they detract from the main activities of the meeting.

I once attended a one-day workshop in which two of the six available hours were expended on establishing ground rules and formulating an agenda. This use of the available time created a great deal of frustration among many of the participants because the time remaining was insufficient to deal with the intended workshop agenda.

Procedures

Meeting procedures can ensure that each group has an equal opportunity to express perceptions and concerns and to have them included in the meeting's generated statements and accounts. By making frequent use of small-group processes, facilitators can provide opportunities for people to articulate their thoughts and ideas in safety. This ensures that multiple perspectives are elicited and that forceful people do not dominate proceedings. In one useful type of small-group process, the facilitator or meeting leader follows these steps:

- Divide the meeting into groups of, usually, no more than six members.
- Describe the activities to be performed or the questions to be discussed (these may be selected from any of the frameworks for developing descriptive accounts presented in this chapter).
- Provide adequate time for these purposes to be achieved.
- Have each group write a summary of activity outcomes on a chart.
- Have all participants reassemble and display their charts.
- Have each group present its summary verbally. As each group presents, questions from the facilitator or audience may allow group members to clarify meanings and in some cases extend their descriptions. This additional information may then be added to the chart.

Decision Making

Meetings should operate on the basis of consensus, rather than on the basis of a majority vote. The latter encourages competitive, divisive politicking, which usually ensures that the least powerful groups will not have their interests met. Although consensus is sometimes difficult to attain, it is a powerful instrument for change when it is achieved. When agendas are pushed through to accommodate time pressures, bureaucratic pressures, or the interests of powerful groups, the outcomes are likely, in the long run, to be unproductive. Time is an essential element in any collaborative activity; it cannot be compressed without damaging the essential participatory nature of a community-based action research process.

Venues

Initial meetings may be held in people's homes, cafes, offices, community centers, or any other venue where the stakeholder group itself is comfortable. When people talk in the comfort of their own territory, they are more likely to be honest and forthcoming. Public venues are appropriate when all stakeholder groups meet to work through issues and agendas. Even then, however, researchers should take care to select meeting sites where the least powerful groups will feel comfortable. If a meeting is held in their territory, they are more likely both to attend and to be willing to contribute to the proceedings. A local community hall, church halls, hostels, community health clinics, lodge halls, and even parks may provide appropriate contexts.

Reflection and Practice

1. Using the techniques presented in this chapter, interview a stranger or someone you don't know very well. You might choose someone from the scene you observed for the previous chapter's activities. Your purpose is to learn how the person experiences that setting or about an important event in the person's life—wedding, family gathering, party, birthday, and so on.
2. Before you start you will need to prepare the following:
   - How you will introduce yourself and the purpose of the interview
   - The type of questions you will ask
   - The method of recording information—notebook or tape
3. After the interview read back the notes you have taken to the person interviewed. Ask whether your notes accurately reflect what he or she said.
4. Reflect on the process and discuss with another person. How did that feel? What did you note about the interview process? How might you have improved? (Keep the notes [data] for later analysis.)
5. Return to the previous setting, or enter another. Observe it for an extended period of time (at least a half hour), using techniques suggested in this chapter. Record what you observe. (Keep these observations for later analysis.)
6. Reflect on the process of observation. How did that feel? What did you learn about the process of observation?
7. From the interview and observation, list what else you would like to learn about that setting? Formulate a short questionnaire around those issues.
8. Return to the setting and interview a number of persons using that questionnaire.
9. Alternatively, prepare a number of copies of the questionnaire and ask people to complete it.
10. Reflect on the differences between the use of a questionnaire and the interview process for gathering information. How did they differ? What were their different strengths? What were their deficiencies?
11. Identify an issue, question, or problem arising in your interviews or observations. Do a search of the research literature on that topic. Identify studies that have been done and summarize what they reveal.

BOX 4.1

Look: Building the Picture

Purpose

To assist stakeholding groups in building a picture that leads to

Understanding: What and how events occur
Clarity: A detailed picture of the context
Insight: An extended understanding of the issue

Process

Gather Information

Sources of information:

Interviews: Key people from each stakeholding group
Observation: Significant settings, events, and/or activities
Other sources: Documents, records, reports, a survey, research literature

Record Information

Record information as

Notes: A written record of what people said and/or did, or key points from documents, records, or the research literature
Audiotapes: Audio record of interviews
Videotapes: Video record of events, people, places, and so on
Photographs: Photographs of events, people, places, and so on

Extend Understanding

Construct frameworks for descriptive accounts

Collaborative descriptive accounts
Six questions—Why, what, how, who, when, where
Community profiles

Organizing Meetings

Participants
Leadership and facilitation
Ground rules and agenda
Procedures
Decision making

Communicating

Inform people of research activities by distributing reports to participants and stakeholders as

Meeting minutes
Bulletins
Interim reports

NOTE

1. Very few action research contexts provide the means to incorporate the necessary research design that would enable a practitioner to rigorously test hypothesis or to generalize the results outside of the research contexts. Transferability to other contexts (Lincoln & Guba, 1985; Stringer, 2004) may be possible if appropriate "thick" descriptions result from the study. The use of "faux" experiments in action research is likely to lead to spurious results (results that cannot be attributed to the "dependent variable" or the "intervention") because of the inability to control variables in an action research setting.

FIVE

THINK

Interpreting and Analyzing

In the first phases of research, participants acquire large quantities of information (data) that must be analyzed. The next task, therefore, is to identify the aspects of the information that will assist people in clarifying and understanding the nature of the activities and events they are investigating. Analysis is the process of distilling large quantities of information to uncover significant features and elements that are embedded in the data. The end result of analysis is a set of concepts and ideas that enable stakeholding participants to understand more clearly the nature of the problematic experiences affecting their lives. These concepts and ideas may then be used to construct reports providing accounts of *what* is happening and *how* it is happening.

Analysis may be envisaged as a process of reflection and interpretation, providing participants and other stakeholding audiences with new ways of thinking about the issues and events investigated. I begin with a brief discussion of the nature of interpretation and then present a selection of frameworks that assist research participants in formulating productive interpretations of issues investigated.

INTERPRETATION: IDENTIFYING KEY CONCEPTS

Denzin (1989) has written of the need to make the problematic, lived experience of ordinary people directly available to policymakers, welfare workers, and other service professionals, so that programs and services can be made more relevant to people's lives. He suggests that an interpretive perspective identifies different definitions of the situation, the assumptions held by various interested parties, and appropriate points of intervention:

> Research of this order can produce meaningful descriptions and interpretations of social process. It can offer explanations of how certain conditions came into existence and persist. Interpretive research can also furnish the basis for realistic proposals concerning the improvement or removal of certain events, or problems. (p. 23)

The task of the research facilitator in this phase of the research process is to interpret and render understandable the problematic experiences being considered. Interpretation builds on description through conceptual frameworks—definitions and frameworks of meaning—that enable participants to make better sense of their experiences. It uses experience-near concepts—terms people use to describe events in their day-to-day lives (rather than, e.g., theoretical concepts from the behavioral sciences)—to clarify and untangle meanings and to help the individuals illuminate and organize their experiences. The researcher must provide the opportunity, in other words, for participants to understand their own experiences in terms that make sense to them.

Interpretive activity exposes the conceptual structures and pragmatic working theories that people use to explain their conduct. The researcher's task is to assist participants in revealing those taken-for-granted meanings and reformulating them into "constructions [that are] improved, matured, expanded, and elaborated" and that enhance their conscious experiencing of the world (Guba & Lincoln, 1989). These new ways of interpreting the situation are not intended as merely intellectualized, rational explanations; rather, they are real-life constructs-in-use that assist people in reshaping actions and behaviors that affect their lives.

"Interpretation is a clarification of meaning. Understanding is the process of interpreting, knowing, and comprehending the meaning that is felt, intended, and expressed by another" (Denzin, 1989, p. 120). The purpose of interpretive work, therefore, is to help participants to "take the attitude of the other" (Mead, 1934), not in a superficial, mechanistic sense but in a way that enables them to understand empathetically the complex and deeply rooted forces that move their lives.

In some instances, initial interpretive work provides the basis for immediate action. Some problems, however, are more intransigent and require extended processes of exploration, analysis, and theorizing. The form of analysis should be appropriate to the problem at hand. Complex or highly abstract theories, when applied to small, localized issues, are likely to drain people's energy and inhibit action. Explanations and interpretations produced in action research processes should be framed in terms that participants use in their everyday lives, rather than in terms derived from the academic disciplines or professional practices.

When I began working at the community level, I would often present explanations for problems as derived from my background in the social sciences. I would include concepts such as social class, racism, power, authority, and so on. In few instances did these forms of analysis strike a chord with the people with whom I was working.

On one occasion, I presented an analysis to a group of Aboriginal people that implied that their responses to racism were inappropriate. My explanation outlined a theory of minority responses to racism that highlighted apathy, avoidance, and aggression as typical behaviors that derived from a history of oppression. The anger directed at me by the Aboriginal people present was a humbling experience.

In retrospect, I understand their anger and wonder at my naiveté. Not only was I interpreting their situation from my perspective; I was also judging their behavior in stereotypical ways, implicitly criticizing their responses to situations they faced. Today, I ensure that any analysis I make is drawn from terms known to the people with whom I am working, is expressed in their language, and is derived from their experience-near concepts. My own role in this process is to assist them in articulating their ideas and to ensure that they express their ideas clearly and accurately. I also ask probing questions that challenge the rigor of their interpretations to ensure that their interpretations will hold up in the court of public scrutiny and future action.

The use of experience-near concepts does not eliminate the need for rigorous inquiry. Restricted or cursory analyses that produce superficial solutions to deep-seated and complex problems are unlikely to be effective. Researchers and facilitators can ensure that explanatory frameworks are sufficiently rigorous to move people past stereotypical or simplistic interpretations of their situations, but these frameworks must be grounded in the reality of their everyday lives. They must acknowledge the experiences and perspectives of those to whom programs and services are directed, rather than of those who deliver those services.

In the early phases of a recent project in East Timor, principals and superintendents often commented that parents were not interested in their children's schooling—that it would be difficult to get them to participate in school life. Through a series of parent workshops I was able to demonstrate that parents had high levels of interest in their children's schooling. Using participatory processes of inquiry based on small-group techniques, these workshops gave parents opportunities to express their concerns about the school and to suggest ways that they might be able to engage in activities that would assist the staff of the school in dealing with the issues that emerged. In most schools parents identified a number of ways that they would be willing to participate in improving the school, and many successful projects emerged—fund-raising; school repair and maintenance; garden maintenance; teaching materials manufacture; local crafts, art, songs, and dances; and assisting with school security arrangements. Parents found many ways that they, with their limited education and resources, could make a difference in their children's schooling.

ANALYSIS AND INTERPRETATION I: DISTILLING THE DATA

Two major processes provide the means to distill the data that emerge from the ongoing processes of investigation. The first is a *categorizing and coding* procedure that identifies units of meaning (experience/perception) within the data and organizes them into a set of categories that typify or summarize the experiences and perspectives of participants. The second data analysis process selects *key experiences* or transformational moments and "unpacks" them to identify the elements that compose them, thus illuminating the nature of those experiences. Researchers may use either or both of these techniques of data analysis as they seek to acquire clarity and understanding by distilling and organizing the information they have gathered.

Categorizing and Coding (1)

The major task of this procedure is to identify the significant features and elements that make up the experience and perception of the people involved in the study (stakeholders). *All* analysis is an act of interpretation, but the major aim in analysis is to identify information that clearly represents the perspective and experience of the stakeholding participants. Those involved in data analysis must "bracket" their own understandings, intuitions, or interpretations as much as possible and focus on the meanings that are inherent in the world of the participants. This is a difficult task that requires some practice and feedback to identify the ways in which we tend to view events through our own perspectives, and it points to the need to "ground" our analysis in participant terms, concepts, and meanings. This is tricky ground, especially when we come to coding procedures, where we must use a term or heading to represent the data within a category:

> Maria Hines, a member of a city neighborhood collective, is most explicit about her experience of analyzing data in a project in which she participated. With a slight frown she describes how "I never knew how difficult it was *not* to put my own words and meanings in. We had to really concentrate to make sure we used what people had actually said and not put it in our own words. It was *hard*." (Stringer, 2004)

To minimize the propensity to conceptualize events through their own interpretive lenses, researchers should, wherever possible, apply the *verbatim principle,* using terms and concepts drawn from the words of the participants themselves. By doing so they are more likely to capture the meanings inherent in people's experience.

Because different stakeholders are likely to have quite different experiences and perspectives on any issue, analysis of each stakeholding group should initially be kept separate, and more general categories developed at later stages of a project. Thus initial analysis will keep, for example, teacher, student, and parent perspectives separate to identify ways that these stakeholders "see" the situation. It will then be possible to understand the elements that their perspectives have in common and the ways in which they diverge. Likewise, city planners, businesspeople, and residents may have differing perceptions of a neighborhood development project that all need to be acknowledged and incorporated into planning procedures.

Procedures for categorizing and coding involve

- Reviewing the collected data
- Unitizing the data
- Categorizing and coding
- Identifying themes
- Organizing a category system
- Developing a report framework

Reviewing the Data

Commence this phase by first reviewing the issue on which the study is focused and any associated research questions. The purpose of analysis is to identify data (information) that is pertinent to these issues and questions. As data analysis continues, there may be considerable amounts of data that are either irrelevant or peripherally relevant, so that choices need to be made about which data to incorporate into processes of analysis.

Researchers should first review transcripts or records of interviews, reading them to familiarize themselves with the contents and to get a feel for the views and ideas being expressed. Other types of information will be incorporated in further cycles of analysis.

Unitizing the Data

As people talk about their experience and perspective, their narrative is composed of a wide range of related and interconnected ideas, activities, and events. They will often change direction or focus on the many parts of the story that compose the interrelated aspects of everyday experience. The next phase is to identify the discrete ideas, concepts, events, and experiences that are incorporated into their description to isolate the elements of which their experience is composed.

Using a photocopy of the original data, block out each separate item of information using a pencil to identify units of meaning. A unit of meaning may be a word, a phrase, a sentence, or a sequence of sentences. You will then literally cut out each of these pieces and paste it onto a card, labeling the card to indicate the origin of that unit—the interview from which it was drawn.

> Teachers and parents came to the next meeting / and talked of starting a PTA (Parent Teacher Association). / Some parents refused to consider this / due to their experience of past associations. / The parents then identified other parents who had good ideas, / who had the interest and enthusiasm, / and who would work with teachers / to improve the school. / They decided they needed to inform other parents / of the possibilities for starting a Parent Teacher Association. . . .

When pasting units onto a card, extra information may need to be added to make the meaning clear. For example, to the unit "Some parents refused to consider this," researchers would need to add in brackets "starting a Parent Teacher Association." The intent is to identify *units of meaning*—statements that have discrete meaning when isolated from other information.

Categorizing and Coding

Once the data has been unitized, the units of meaning must be sorted into related groups or categories. The previous boxed example would provide information about parent activities, experiences, and perspectives about a PTA. It would provide the basis for a category that may be identified (coded) as "Parents' Perspectives on a PTA." Information from other interviews would be added and the category resorted to identify different dimensions of their perspective. Each pile of cards, or category, would then be labeled to identify the particular dimension; for example, an initial set of categories based on the previously discussed data might be "Starting a PTA," "Parents With Good Ideas," or "Parents With Interest and Enthusiasm." As data from other interviews are included, however, these categories might be seen as inappropriate and the code revised.

As the data are analyzed, categories might emerge that enable a large number of activities to be included under a relatively small number of headings. For instance, analysis of interviews that focus on PTAs might reveal the following categories: "Organizing a PTA," "The Structure of a PTA," "Parent Activities," "Improving the School," and so on. This is quite a natural way of organizing information. In everyday life we would include oranges, apples, pears, and peaches within a category called "fruit." Likewise, shirts, shorts, slacks, and sweaters might be categorized as "clothing."

Identifying Themes

When the categories associated with each stakeholding group have been placed in a system of categories it may be possible to identify themes that are held in common across stakeholder groups. Within a school we may see that teachers, students, and parents are concerned about "results," even though their concerns are expressed differently. Neighborhood stakeholders, similarly, may be concerned about the effect of a new roadway, though they may see that effect in either positive or negative ways. All perspectives need to be incorporated under the overarching theme "Effect of a New Roadway."

Research participants therefore need to identify themes—issues or experiences that people have in common—by comparing categories and subcategories across stakeholding groups.

Organizing a Category System

The category system must then be recorded in some rational form, providing a clear picture of the categories and subcategories of information related to the topic investigated.

The list of contents for this chapter provides an example of a category system. The broad theme "Think: Interpreting and Analyzing" presents a number of major features of analysis. Next, the contents list elements of each of these followed by subelements. Thus the features and elements of the process of analysis are organized in a logical order that assists readers in understanding the material being presented:

Think: Interpreting and Analyzing

Interpretation: Identifying Key Concepts

Analysis and Interpretation I: Distilling the Data

Categorizing and Coding (1)

Analyzing Key Experiences (2)

Case Example: Facilitating Workshops

Analysis and Interpretation II: Enriching the Analysis

Extending Understanding: Frameworks for Interpretation

Alternative 1: Interpretive Questions: Why, What, How, Who, Where, When

Alternative 2: Organizational Review

Developing a Report Framework

This structure provides a framework for reports or presentations that communicate the outcomes of this phase of the research to relevant stakeholders. Themes, categories, and subcategories provide the headings and subheadings for this purpose.

Analyzing Key Experiences (2)

The purpose of this approach to analysis is to focus on events that seem to have a marked impact on the experience of major stakeholders. Denzin (1989) talks of moments of crisis, or turning-point experiences that have a significant impact on people. Such events may appear as moments of crisis, triumph, anger, confrontation, love, warmth, or despair that have a lasting impact on people. They may result in a "lightbulb" or "aha" experience that provides people with greater clarity about puzzling events or phenomena, or leave them with deep-seated feelings of alienation, distrust, affiliation, or hope.

As we interview people, especially about issues in their lives, they are likely to focus on events that have special significance for them. By "unpacking" these events we can learn the features of that experience that make them so meaningful, and in the process we extend our understanding of the way the issues affect their lives.

Review the Data

Review the data as suggested in "Categorizing and Coding."

Identify Key Experiences

For each participant, identify *events or experiences* that appear to be particularly significant or to have an especially meaningful impact on them.

Identify Main Features of Each Experience

For each significant event or experience, identify the *features* that seem to be a major part of that experience.

Identify the Elements That Compose the Experience

For each feature, identify the elements that compose the detailed aspects of that experience.

Identify Themes

List experiences, features, and elements for each participant. Compare lists to identify experiences and features of experience that are common to groups of participants. List these as themes.

Case Example: Facilitating Workshops

The following case provides an example of part of a study on a group of people whose work involved facilitating workshops. The analysis commences with a piece of raw data and demonstrates how the analysis identifies a key experience and its associated features and elements.

> I remember the anxiety that used to sit in the pit of my stomach as I watched my colleagues commence a workshop, my mind racing with details of what needed to happen and the "what-ifs" that haunted me. My organizational mind was constantly working on plans to institute if things failed to eventuate as they were intended. It was something that nagged me and caused me many sleepless nights.

I eventually learned that the start of a workshop was always likely to be a bit "lumpy" and that, provided facilitators were well prepared and flexible enough to accommodate the needs and capabilities of participants, we would almost always accomplish our goals.

So now I can watch my colleagues facilitating a workshop and appreciate the careful planning that has occurred and sit proudly as they work competently with the participants, engaging their attention and enthusiasm. I watch the energy of the participants as they focus on the real issues emerging from their work. I feel that I have become blessedly redundant and that my colleagues have the capacity, evolved from their previous work with me, to carry on the business at hand. They're clear, competent, and well-spoken, able to achieve excellent results and to enable participants to achieve the solutions they have sought through their participation.

Key Experience

Experiencing an effective workshop

Features and Elements

- Anxiety
  - Feelings
  - Sleeplessness
  - "Lumpy" start to workshop
- Careful planning
  - Identify outcomes
- Identify learning tasks
  - Sequence tasks
  - Report
  - Evaluate
- Flexible processes
  - Explain each task
  - Demonstrate or model the activity
  - Participants enact the activity
  - Debrief or feedback (evaluation)

Repeat if necessary
Timing according to participant needs, skills, or understanding

Accomplishing purposes
Engaging attention and enthusiasm
Focusing on real issues
Working competently
Having the capacity
Achieving solutions
Feeling proud, redundant

This example provides the key features "Anxiety," "Careful Planning," "Identifying Learning Tasks," "Flexible Processes," and "Accomplishing Purposes." The elements composing details of the first and last features are drawn from the previous data. A following interview revealed the (italicized) elements composing "Careful Planning" and "Flexible Processes." These features and elements would provide the basis for writing a report that enables people to understand how I experienced an effective workshop.

ANALYSIS AND INTERPRETATION II: ENRICHING THE ANALYSIS

The first cycles of an action research process enable researchers to refine their focus of investigation and to understand the ways in which primary stakeholders experience and interpret emerging issues. In following cycles other information is incorporated that further clarifies or extends participants' understandings by adding information from other stakeholders and data sources. In a school research process, the perspectives of parents might be added to those of students and teachers, and school or student records or the research literature might provide relevant information. In a health program, patient and health professional perspectives might be complemented by evidence-based information from the professional literature.

The purpose for this activity is to provide the means for achieving a holistic analysis that incorporates all factors likely to have an impact on

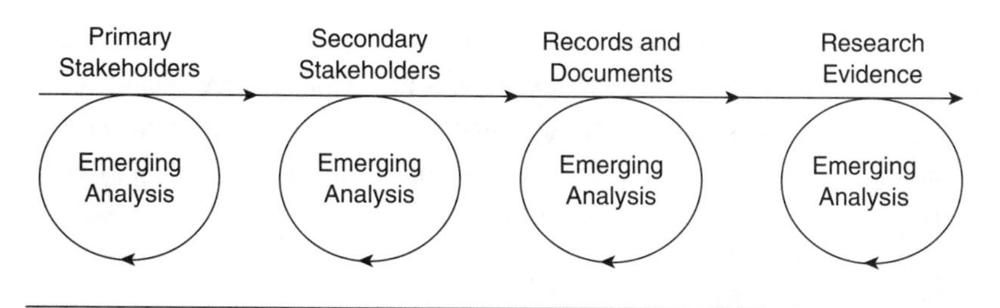

Figure 5.1 Incorporating Diverse Data Into an Emerging Analysis

achieving an effective solution to the problem investigated. Thus the part that each major stakeholder plays will be taken into account, as well as the substantial information that exists in policy and program documents and the research literature.

The example previously presented of parent concerns for mixed-grade classrooms provides an example of this process. Parents were able to clearly articulate their desire to ensure that their children's education would not suffer from these arrangements, and information from the research literature provided them with reassurance. The diagram in Figure 5.1 presents ways that different types of information may be incorporated into cycles of an action research process.

EXTENDING UNDERSTANDING: FRAMEWORKS FOR INTERPRETATION

When people engage the significant issues embedded in long-standing or highly contentious problems, they sometimes need to engage in extensive research to identify the underlying issues involved in the situation. Frameworks of analysis that enable them to delve beneath the surface of events can provide rich understandings that enable them to "make sense" of the situation and deal effectively with the complex problems that often are part of their experience. Four alternative approaches or frameworks that assist stakeholders in this process of sense making and clarification are presented in the following sections: *interpretive questions, organizational review, concept mapping,* and *problem analysis.* Larger or complex projects may require more extended and detailed frameworks, such as those found in the literature on management, planning, community development, and/or applied anthropology (e.g., Black, 1991; Van Willigen, 1993; Whyte, 1984).

Alternative 1: Interpretive Questions—Why, What, How, Who, Where, When

For relatively simple projects, participants might work through a series of questions that enable them to extend their understanding of the problems and contexts they have previously described. Interpretive statements that result from this process should help them develop accounts that reveal the nature of the problem at hand and important features of the context that sustain it. Interpretive questions might include the following:

- Why are we meeting? (purpose, focus, problem)
- What are the key elements and features of the problem?
- How is the problem affecting us? What is happening?
- Who is being affected?
- Where are they being affected?
- When are they being affected?

It may be useful for the facilitator to repeat each of these questions a number of times to enable participants to build more information into their interpretations.

More complex projects may require more detailed and extended questioning strategies, to reveal such elements as the following:

- The history of the situation—how it came about
- The individuals, groups, and types of actors involved
- Interactions and/or relationships among the people involved
- The purposes and intents of the people involved
- The sequence and duration of related events and activities
- The attitudes and values of the people involved
- The availability of and access to resources, and their use

The following types of questions may be relevant:

- Who?
  - Who is centrally involved? (individuals, groups)
  - Who else is significant?
  - Who are the influential people? What is the nature of their influence?
  - Who is linked to whom? In what ways?
  - Who is friendly or cooperative? With whom?
  - Who is antagonistic or uncooperative? With whom?
  - Who has resources? Which ones?
- What?
  - What major activities, events, or issues are relevant to the problem?
  - What is each person or group doing, or not doing?
  - What are their interests and concerns?
  - What are their purposes? What do they want to achieve? What do they want to happen?
  - What do they value?
  - What resources are available? (people, material, equipment, space, time, funds)
  - What resources are being used? By whom?
- How?
  - How do acts, activities, and events happen?
  - How are decisions made?
  - How are resources used?
  - How are individuals and/or groups related to each other?
  - How do different individuals and groups affect the situation?
  - How much influence do they have?
- Where?
  - Where do people live, meet, work, and interact?
  - Where are resources located?
  - Where do activities and events happen?
- When?
  - When do things happen?
  - When are resources available?
  - When do people meet, work, and engage in activities?
  - When does the problem occur?
  - What is the duration of occurrences of the problem?

Not all these questions will be relevant to any one situation. The questions the facilitator selects will depend on the context of the problem and the setting. The researcher should record in detail answers to the questions and should write a summary on a chart that can be seen by all members of the group. Such charts help participants visualize the situation they are interpreting and provide records that can be employed in subsequent activities.

Alternative 2: Organizational Review

In some circumstances—where, for instance, different sections of the same institution or agency experience similar problems—it may be appropriate for the researcher to conduct a review of the whole organization. This activity is intended to reveal different interpretations of problematic features of the organization and sources of the problems. It is intended not as an evaluation of competence or assessment of performance but as a method for discovering points where action can be taken. Participants in the review process should focus on the following features of the organization:

Vision and Mission

- *Vision:* What is the overarching or general purpose of the organization? Education? Health improvement? Assistance for the needy?
- *Mission:* In which ways does the organization seek to enact its vision? Providing educational services, courses, and classes? Engaging in health promotion programs? Providing welfare services?

Goals and Objectives

- *Goals:* How does the organization seek to achieve its purposes? In what activities does it engage?
- *Objectives:* What are the desired outcomes of these activities? For whom?

Structure of the Organization

- *Roles:* Are roles clearly delineated? Who works with whom? Who has authority over whom? Who supervises and gives directions? To whom?

- *Responsibilities:* What types of people are responsible for different categories of activities? Who performs which types of tasks?
- *Rules and procedures:* Is it clear what needs to be done and how it is to be done?
- *Resources:* Are the resources required for tasks adequate and available (e.g., time, materials, and skills)?

Operation

- Is each person clear about his or her roles and responsibilities?
- How effectively is each person enacting his or her roles and responsibilities?
- What factors hinder the enacting of those responsibilities (e.g., lack of materials, time, skills, or support)?
- Are there tasks and responsibilities to which no one is clearly assigned?
- What is not happening that should be? What is happening that should not be? Where are the gaps? Where are the barriers?

Problems, Issues, and Concerns

- What problems, issues, and concerns are expressed by stakeholders?
- Who is associated with each?
- How do stakeholders explain or interpret problems, issues, and concerns?

As participants work through these issues, they will extend their understanding of the organization and aspects of its operation that are relevant to their problems, issues, and concerns. The outcomes of their inquiries should be recorded in detail and summarized on charts for use in later stages of the research process.

I once worked as a consultant with a community child care agency whose mandate was to find appropriate foster homes for children and to ensure that the foster families were adequately supported. The agency was experiencing a number of difficulties in performing these functions. Staff members were stressed and overworked, and funding sources questioned the effectiveness of the agency's services. When we charted the purposes for which the agency had been formed and matched those against its actual activities, we discovered that staff were engaged in work that was of peripheral relevance to the agency's stated main function. They were attempting to provide a wide range of support services to foster families—furniture, clothing, food, and counseling—that could be provided by other agencies funded for those purposes. When they saw what was happening, staff and committee members decided to terminate many of those activities and to link clients to other relevant agencies when such needs arose.

Alternative 3: Concept Mapping

Long-term, entrenched problems frequently defy the remedial efforts of professional practitioners. Occurrences of drug abuse, eating disorders, lack of interest in schoolwork, youth crime, and other problems are so pervasive that they often appear beyond the reach of ordinary programs and services. They exist as part of a complex system of interwoven events and circumstances that are deeply embedded in the social fabric of the community.

Attempts to remediate these problems are unlikely to be successful if they focus merely on one aspect of the interrelated factors that make up the situation. Quick-fix, "spray-on" solutions rarely work. Participants need to reconceptualize such issues in ways that clearly identify the interrelationships among all the significant elements that affect the situation. The intent of this type of activity is to help participants find a way to pursue multiple, holistic, and inclusive strategies that will assist people in dealing with the problems that affect their lives.

In the first phase of the research process, delineated in Chapter 4, participants describe the problem and define key elements or characteristics of the situation. In concept mapping, those elements are plotted diagrammatically, so that participants can visualize the ways in which different components of the situation relate to the problems they are investigating. To guide participants in formulating a concept map, the facilitator should take the following steps:

- Begin by printing on a large piece of paper or a board visible to all participants a word or phrase that characterizes the central problem, and then enclose the word or phrase inside a geometric figure (e.g., a square or circle).
- Add to the chart other geometric figures labeled to represent various elements associated with the problem.
- Link the figures containing the elements that seem to be related to each other.
- Extend the mapping process to include additional figures until each participant is satisfied that all significant elements have been included.

An example of a concept map is provided in Figure 5.2. This interpretation of poor school behavior of youth in a rural town is ascribed by stakeholders to a number of interlinked factors—poor attendance, low academic attainment, general family issues, parent issues, financial problems, and poor health. All issues are seen to contribute to the identified behavior problem and need to be addressed as part of the process of defining a solution to the problem of poor school behavior. Further analysis may have extended this map to delineate factors associated with health problems, financial problems, and so on. Financial problems, for instance, may be associated with high unemployment in the town, which in turn may relate to closure of businesses or other economic conditions.

Concept maps help participants visualize the major influences that need to be taken into account and assist them in evaluating whether all relevant stakeholders have been included in the investigation. In Figure 5.2, the appearance of family issues signals the need to involve relevant family members in the inquiry process. The young people who are themselves the focus of the

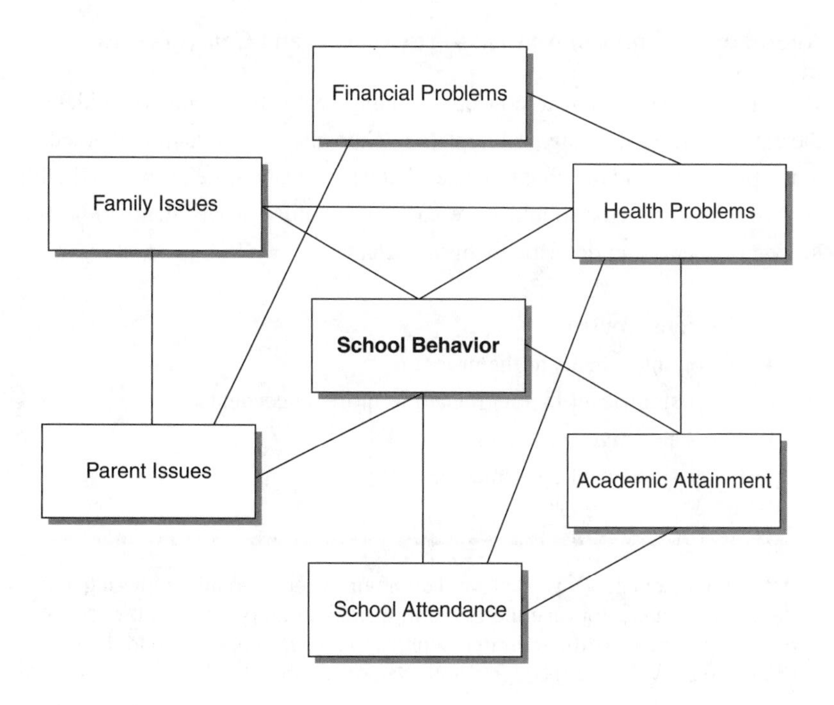

Figure 5.2 Concept Map of Poor School Behavior

problem should be central to the concept-mapping process. Incorporating additional stakeholders in turn may extend the factors incorporated into the map and thus the elements that need to be considered when action is planned. Activities from other interpretive frameworks (e.g., interpretive questions) can be used to further expand this form of analysis.[1]

The explanatory frameworks produced in this process are not meant to be scientifically or professionally defensible. We are looking not for the best or the correct explanation but for one that makes sense to or can be accommodated by all the stakeholders. In all the processes described earlier, it is essential that each stakeholding group provide input about its own situation. Analyses based on other people's interpretations do not provide an appropriate basis for action. In Figure 5.2, for instance, teacher or administrator perspectives on the family are less relevant than the perspectives of the family members themselves. Outsider interpretations often are incomplete or inaccurate and are sometimes judgmental; they may tend to create divisiveness and hostility, which are antithetical to the participatory principles of community-based action research.

Alternative 4: Problem Analysis—Antecedents and Consequences

An approach similar to concept mapping enables participants to identify the antecedents of existing problems (i.e., elements of the situation that led up to the problems) and the consequences that derive from those problems. In this process, the facilitator should have each stakeholder group identify the following elements and describe them on a chart:

- The core problem
- Major antecedents to the problem
- Other significant factors related to those antecedents
- Major negative consequences
- Other significant consequences

A committee of agency workers and community representatives investigated the high incidence of juvenile crime in a small country town. A process of exploration revealed that criminal activity appeared to be related to the lack of leisure activities for teenagers and to poor school attendance. No one in town accepted responsibility for organizing activities, there were few facilities available for that purpose, and the school provided little in the way of after-school programs or activities. Poor school attendance was directly related to poor school performance, uninteresting or irrelevant curricula, and family poverty.

As a result of increases in juvenile crime, burglary, petty theft, and vandalism there was a general increase in alienation of some groups of young people from other sectors of the community. There were increasing incidents of conflict between youth and members of the community, businesspeople, law enforcement personnel, teachers, and so on, and deterioration in employment opportunities for young people. The long-term prospects of some youth also seemed likely to be affected by the negative outcomes of incarceration in prisons and other institutions and growing criminal records. This situation is drawn as a concept map in Figure 5.3.

This process is sometimes defined as *causes* and *effects* rather than antecedents and consequences. In many situations, however, it is impossible to delineate the difference between cause and effect because of the complexity of interactions between the constituent parts of the picture. The same phenomenon often can be envisaged as cause, effect, or both. Nevertheless, this approach can help participants identify significant elements of their analysis and visualize aspects of the situation that require action.

WRITING REPORTS COLLABORATIVELY

Researchers commonly work independently to analyze data and formulate reports on the basis of their own interpretive lenses. In the process they tend to lose the interpretive perspective of other participants and to fashion a report that often fails to adequately represent the perspective of those with whom they work. The following procedures provide the means to engage stakeholding groups in the processes of analysis, thus ensuring that the end result integrates their perspectives and priorities. Research facilitators help participants engage in meaning-making discussion and dialogue—hermeneutic dialectic processes—with the intent of developing mutually acceptable accounts of the issues and problems they are investigating. To facilitate these processes, researchers assist selected participants with organizing meetings, setting an agenda, reviewing descriptive accounts, and analyzing information and developing accounts and reports.

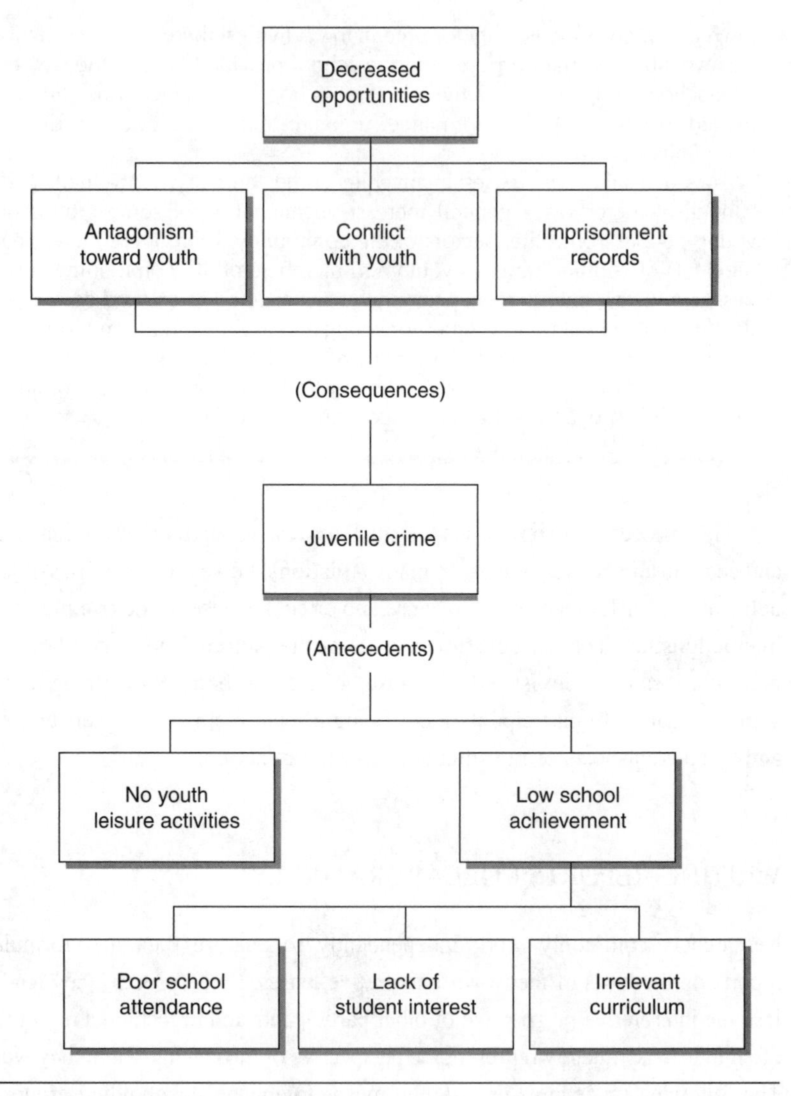

Figure 5.3 Antecedents and Consequences of Juvenile Crime

Organizing Meetings

Research facilitators will work with key stakeholders to organize meetings (see Chapter 4). It is important that members of the principal stakeholding group, for whom the problem or issue is most central, have a key role in the organization of meetings. These members should ensure that representatives who can speak for each of the stakeholding groups are able to attend.

Procedures for Analysis

Setting the Agenda

In this phase of the process, participants learn something about the *people* involved, the *purpose* of meeting, and the *activities* in which they will be engaged. The facilitator's role includes the following duties:

- Inform people of the purpose of the meeting.
- Provide an opportunity for participants to introduce themselves and identify the groups to which they belong.
- Present a broad agenda for the session.
- Allow time for participants to discuss, clarify, and modify the agenda. Do not discuss the *issues* at this time, but focus on the processes of the meeting.

Reviewing Descriptive Information

The purpose of this activity is to give participants a chance to review the descriptive information produced in previous meetings. This is especially important when stakeholding groups have done some initial work separately. Facilitators should provide opportunities for each stakeholding group to do the following:

- Send written versions of the descriptive accounts or give them to participants prior to the meeting.
- Present a verbal report that summarizes descriptive information produced by their previous activities.
- Present the key elements of these reports on a chart.
- Allow time for meeting participants to discuss and clarify the accounts, avoiding extended or detailed discussion unless there are contentious issues that need to be resolved.

Distilling the Information: Analysis

Using one of the processes for data analysis presented earlier, participants should work collectively to organize the information in the charted summaries into sets of categories. They should identify *converging perspectives* (i.e., those ideas, concepts, or elements common to all or a number of groups) and *diverging perspectives* (i.e., those ideas, concepts, or elements found in the accounts of only one or a few groups). Because this activity is time-consuming, it may sometimes be appropriate to formulate a *working party* to carry it out:

- Participants should select elements from the charts that are perceived to be significant features or characteristics of the issue investigated.

- Participants should sort elements into categories, so that those associated in some significant way are clustered together. The idea is to rationalize the large number of individual ideas, accounts, or issues to create a smaller number of categories. In doing so, facilitators should ensure that the process of categorizing does not erase important information. A category labeled youth issues, for instance, might erase important distinctions between early and late teens, or between males and females. Decisions about ideas that can be incorporated into particular categories and those that must remain separate should be made through discussion and negotiation.

- Participants may organize categories in a variety of ways. A color-coding system may be used, for example, wherein similar items are identified on charts with colored markers. Alternatively, each idea, issue, or concept may be copied onto a card, and the cards can then be sorted into piles according to their similarities or common characteristics. Groups may also use large-sized self-adhesive labels, placing all elements on a large wall and allowing participants to move them around into groups associated together. Participants should know that there is no right way or correct answer for this process. The way that they organize the information is the way that makes best sense to them collectively.

- When participants have reached agreement on the categories and the information contained in each, they should classify each group of concepts according to a label or term that both identifies and describes the category. This process is represented symbolically in Figure 5.4.

Constructing Reports

The previously described process provides material that forms the basis for reports and identifies actions to be taken by participants. Research facilitators

Identified Categories

CHILDREN

TEACHERS

PLAY

SCHOOL

Figure 5.4 Example: Issues Derived From Categorized Information

should continue to meet with participating stakeholders—a working group to formulate a report. The working group should

- Review materials
- Use the category system developed during analysis as a framework for writing a report
- Use the outline to write a detailed written report (Detailed records from meetings should provide the content of the report; see sections on reporting in previous chapters and in Chapter 8.)
- Provide the time and opportunity for all members of the working group to read the report and give feedback
- Modify the report on the basis of their comments
- Meet again with the working party if any members suggest significant revisions
- Distribute the report to all stakeholders

Presentations and Performances

In his book *Interpretive Ethnography,* Denzin (1997) signals the need for forms of reporting that more clearly represent people's lived experience. Current formats tend to render people's lives in forms that are stiff, formalistic, and often encompassed in technical language. They silence the voices of the people to whom they refer and mask the realities of their day-to-day lives. Denzin suggests the need for reports that are more evocative, enabling readers to more readily understand people's experiences. Such reports would be open-ended, multifaceted, and multivoiced and would not indulge in abstract terminology or privilege the perspective of persons in positions of authority. Thus he suggests the need for writing reports that experiment with genre, voice, and narrative style, so that *official* reports may take on the appearance of writing more usually associated with fiction or poetry.

He extends this argument to suggest that such writing may be envisaged as *texts waiting to be staged.* That is, people may be better able to represent their ideas and experiences through performances than through written reports. Thus it may be possible to present the outcomes of a research process through

- Drama
- Role plays
- Simulation
- Dance
- Song
- Poetry
- Works of art
- A combination of the above

This move to performance is especially important in action research because participants often include people whose feelings of well-being are diminished by the intrusion of formalistic texts. Researchers need to find ways of communicating that do not disempower groups of participants.

A researcher once showed me a report she had written for a government agency that presented an Aboriginal community's account of a research process in which they had engaged. I asked her how many people in the community had read the report, and she admitted that few would wish or be able to read what was rather a thick and formal report. "How," I asked her, "will people get to know about the information in the report?" Her first impulse was to write another version of the report, but I suggested she think of a concert. This would enable different groups to present their perspectives in ways that were more in keeping with the life of the community. Thus people could choose to use song, dance, drama, poetry, humor, role plays, or other forms of presentation to get their message across.

Modern technology opens up possibilities for communication that were not available in the past. Groups may choose, for instance, to use carefully scripted video presentations to present information in their own voice and in their own way (Schouten & Watling, 1997). This form of presentation provides many dramatic and effective possibilities for communicating the outcomes of a research process.

I recently participated in a departmental review process whereby each of the units of the department reported on their activities for the year. The administration section staff, largely younger women, were somewhat shy and loath to present personally, so they produced a video. Taped in their offices, in a swimming pool, and in the garden, all members of the administrative staff reported on their activities. Their presentations were often touched with humor, as was the case of one woman who would speak to the camera only through a hole in a paper bag over her head. The video presentation was engaging and informative and provided other staff with a clear understanding of the work in which administrative staff were engaged and the issues that confronted them.

There are contexts and purposes, therefore, that may be better served by more creative thinking about the best ways for communicating information to audiences of stakeholders. Although formal written reports may be useful in some contexts, it is possible that more innovative uses of narrative texts, staged performances, or electronic productions may enhance the work in which we are engaged. Researchers need to keep the audience and the purpose clearly in mind as we formulate ways to communicate effectively the outcomes of our research processes.

CONCLUSION

The previously described procedures provide the means by which people can formulate clear, sophisticated, useful explanations and interpretations of their situations. The specific ideas and concepts contained within these interpretive frameworks provide the basis for planning concrete actions to remedy the problems on which the research has focused.

For complex problems involving multiple stakeholder groups, the activities described in this chapter may be enacted as separate parts of an action research process. For simpler problems within discrete settings, such as classrooms, offices, and small organizations, they may be incorporated into processes that move more directly from description, through interpretation, to problem solving. It is well to keep the distinction between these activities in mind, however, to ensure that people are clear about the nature and purpose of the processes in which they are engaged.

As presented here, the procedures may seem long-winded and complex, but experience will show researchers when and how to consolidate or abbreviate them to ensure a brisk, purposeful flow of activity. Researchers should be wary, however, of simplifying the research process by confining it to a small inner circle or by omitting "troublesome" stakeholders. Participation boosts personal investment in the process, extends people's understanding of the contexts and social processes in which they are involved, and minimizes the possibility that the research will bog down in conflict. Action research is not just a tool for solving problems, but is a valuable resource for building a sense of community.

Reflection and Practice

Using the field notes gathered in interviews for previous activities,

1. Analyze the information using a *coding and categorizing* process. (If you have extensive notes, you may focus on one issue that emerged within the interview).
2. Present the resulting framework of key features and elements to a friend, colleague, or working group.
3. Explain how that framework will enable you to provide an account of the person you interviewed.

4. Reflect on and discuss the process of analysis. What did it feel like? What are some of your concerns or questions about the process? How was it informative or helpful?
5. Repeat that process using the technique for *analyzing key experiences.*
6. Reflect on and discuss that process. How were the outcomes of the two processes different?
7. Using features and elements from one of the frameworks, draw a concept map. Use one issue as the central element of the map.
8. Using features and elements from each framework, present short reports to your friend/colleague/classmate.
9. Reflect on your presentations. Are there more effective ways you might have presented that information using some of the techniques suggested in the section on Presentations and Performances?
10. Analyze the notes from your observations, using a coding and categorizing process. Write a short report about that setting.

NOTE

1. Deshler (1990) provides a more extensive treatment of concept mapping.

BOX 5.1

Think: Interpreting and Analyzing

Purpose

To use interpretive processes to distill the information gathered, identify key features and elements of people's experience, and enable participants to understand more clearly the way the issue affects their lives and activities.

Process

Frameworks for Analysis and Interpretation

Categorizing and Coding

Review and unitize the data.
Formulate categories and code them.
Identify themes across the categories.
Organize a category system.
Develop a report framework using the category system.

Analyzing Key Experiences

Review the data and identify key experiences.
Identify main features and elements of each experience.
Identify major themes across experiences.
Organize a report framework using features and elements.

Enriching Analysis Using Frameworks for Interpretation

Interpretive questions: Why, what, how, who, where, when
Organizational review: Vision, mission, structure, operation, problems
Concept mapping: Issue, concepts/influences, links
Problem analysis: Problem, antecedents, consequences

Writing Reports Collaboratively

Organizing meetings
Procedures for analysis
- Setting the agenda
- Reviewing information
- Distilling information—analysis
- Constructing reports

Presentations and Performances

Use drama, song, dance, art, and/or poetry for effective presentations

SIX

ACT

Resolving Problems—Planning and Implementing Sustainable Solutions

FROM PROBLEMS TO SOLUTIONS

Chapter 5 described how analysis enables participants to identify the problematic features and elements of the issue on which research is focused. In the next phase of action research, participants will work creatively to formulate actions that lead to a resolution of the problem(s). What can we do, they should ask, that will enable us to achieve better results, or a more positive outcome? What steps can we take to ensure that we accomplish the outcomes we desire? Elements and categories that have emerged from the interpretive processes described in the previous chapter suggest key areas or aspects of the situation that need to be dealt with in any plan for taking action. Participants then work creatively to identify *what* they will do to gain a more positive outcome, and *how* they will go about the tasks they have set themselves. This chapter therefore presents systematic ways of planning and implementing the action plans that distinguish action research from other approaches to inquiry.

I've seen superb results achieved through action research processes. In classrooms I've seen students, previously unengaged and uninspired, excitedly and animatedly focusing on their work. I've seen teachers, exasperated with the apparently inflexible nature of their context, find concrete and productive ways of reframing their classroom activities. In one school a single interview was sufficient to provide a teacher with the time to run quickly through a "look-think-act" routine, a quite unintended, but palpably productive process. She described the situation, subjected it to analysis, and formulated solutions almost instantaneously.

"Dr. Stringer, I don't know what I can do about parent-teacher conferences. We only have ten minutes with the parents for each child, and that's only time to go through the report, and time's almost up. We can't really engage in any decent conversation about the child's progress or any problems we're experiencing." [Look] As she extrapolated on these comments she started to explore the issue further. "It's very frustrating, because parents can really be a great help when they work with us. What I really need to do is to spend time explaining how I set up lessons, what I'm trying to achieve with their child, and the problems they are having. They'd then understand what's happening and might be able then to help their child do what needs to be done to improve their work." [Think] She then started to think of ways that she could achieve these purposes. "You know it's impossible to do all that in the time available. What I could do is to meet with groups of parents and explain my teaching processes and problems that their children experience. We could then talk about how parents could do some things to assist with this, or provide their children with specific ways of helping them. We wouldn't need to . . . " She then explored a number of ideas related to this and left, bubbling with excitement and enthusiasm. Although I left that city shortly after, I heard later that she and other teachers had instituted a number of ways to improve parent-teacher conferences and that the process had proved so productive that the school had allocated more time to these activities.

Although all problems are not as easily resolved, it now never surprises me when people are able to see how to modify and adapt their activities and come up with effective solutions to the problems they are experiencing. The process of action research has, for me, high levels of success.

People often assume that a professional analysis provides the best way of envisioning a problem and that all that is then needed is to provide a "recipe" or prescription that people can follow. The problem with generalized recipe-like solutions is that they fail to take account of the underlying issues that have made the experience problematic for participants in the first place. Recipe-based solutions often are based on the professional expertise of the practitioner and fail to take account of the deep understandings that people have of their own experience and the underlying issues that are a central part of the problem.

Professional knowledge can only ever be a partial and incomplete analysis of the situation and needs to complement and be complemented by the knowledge inherent in participant perspectives.

Participants, especially the primary stakeholders whose issues are the central focus of the research, are therefore engaged in further processes of inquiry that provide them with opportunities to conceive of solutions to problems. They then formulate plans that enable them to systematically enact the required tasks and activities. The following procedures are based on a framework of action that involves three phases:

- Planning, which involves setting priorities and defining tasks
- Implementing activities that help participants accomplish their tasks
- Reviewing, in which participants evaluate their progress

PLANNING

In the planning phase, research facilitators meet with major stakeholders to devise actions to be taken. As stakeholders devise a course of action that "makes sense" to them and engage in activities that they see as purposeful and productive, they are likely to invest considerable time and energy in research activities, developing a sense of ownership that maximizes the likelihood of success.

Identifying Priorities for Action

Sometimes the results of analysis identify a single issue needing attention, and research participants may formulate a plan immediately. Often, however, there are multiple related issues or a number of subsidiary issues requiring action, so participants will need to make decisions about the issue on which they will first focus and some order of priority for other issues.

To accomplish this, participants should

- Identify the major issue(s) on which their investigation focused
- Review other concerns and issues and that emerged from their analysis
- Organize the issues in order of importance
- Rate the issues according to degree of difficulty (it is often best to commence with activities that are likely to be successful)

- Choose the issue(s) they will work on first
- Rank the rest in order of priority

Action Plans

Participants then plan a series of steps or tasks that will enable them to achieve a resolution of the issue(s) investigated. Each issue is first restated as a goal. For example, the issue of increasing juvenile crime might be restated as a goal: "to decrease juvenile crime." The features and elements related to this issue, revealed through the processes of analysis, are restated as objectives. The analysis of juvenile crime depicted in Figure 5.3 (in Chapter 5), for instance, suggests that features related to increased juvenile crime include "lack of youth leisure activities," "poor school attendance," and "lack of after-school programs." These could be stated as a set of objectives: "to develop youth leisure activities," "to improve youth school attendance," and "to organize after-school programs." Teams of relevant stakeholders should develop a plan for each issue and bring them to plenary sessions for discussion, modification (if necessary), and endorsement.

A related set of outcome statements provides the means for evaluating the success of the planned activities. In terms of the previously mentioned objectives, outcomes statements would describe the outcomes more clearly—for example, how many leisure activities have been organized, and how well attended, within a certain timeline; what is the specific extent to which school attendance is to be improved; and what type of after-school programs were there, with what attendance, within which timeline.

It is essential that each planning group include a member from each of the major stakeholding groups. The issue "poor school attendance," for instance, should have a team that includes teachers, youth, school administrators, and parents. A simple six-question framework—why, what, how, who, where, and when—provides the basis for planning:

- Goal (Why): State the purpose of the project—for example, to combat juvenile crime. (This can be defined as a goal statement that describes the broad issue to be addressed.)
- Objective (What): State what actions are to be taken as a set of objectives—for example, to organize an after-school program for teenagers and to develop a youth center.

- Tasks (How): Define a sequence of tasks and activities for each objective. List them step-by-step.
- Persons (Who): List those responsible for each task and activity.
- Place (Where): State where the tasks will be done.
- Timeline (When): State when work on each task should commence and when it should be completed.
- Resources: List resources required to accomplish the tasks. Include funds needed to pay for materials or services, or list these separately.

Action plans should be recorded on a chart or whiteboard so that people can get a clear picture of how they will achieve their goal(s) (see Figure 6.1). They provide a concrete vision of the active community of which they are a part and enable participants to check on their progress as they work through the various stages of the project together. Once planning is completed, participants can check that each issue has an action plan and that each person is clear about his or her responsibilities. They can also check the availability of material and human resources required for the tasks they must complete.

Outcome Statements

Outcome statements describe what will have been actually achieved—the end result of the activities delineated in the plan. Where objective statements present a list of intentions, outcome statements describe what is actually going to be done. The following examples map out the outcomes for the Youth Center and After-School Program plans.

Youth Center Outcomes

By June 2, 2008, the Youth Center team will have

- Obtained permission from the principal and the school board to use the disused independent classroom for a Youth Center
- Completed repairs to flooring and repainted the classroom
- Formulated a plan for the operation of the Youth Center as a venue for after-school activities

After-School Program Outcomes

1. By July 3, 2008, the After-School Program team will have
   a. Established an art program

Goal (**Why:** What are we trying to achieve?)
To decrease youth crime by providing an after-school program of activities

Objective *What*	*Tasks* *How*	*Person(s)* *Who*	*Start* *When*	*Finish*	*Location* *Where*	*Resources*	*Funds*
1. Establish a youth center.	a. Obtain permission to use old school building. b. Repair and renovate. c. Plan organization of sports center.	Mrs. Farole Parent volunteers Students volunteers	3/4/08	6/2/08	Disused school building	Paint Timber Tools	$15,000 School district Village council Government grant
2. Establish an after-school program.	a. Establish an art program. b. Establish a sports program. c. Establish a tutoring program.	Mr. Baldwin Jose Marina Venus	4/1/08	7/3/08	Sports Center	Art materials Sports equipment	$3,500 School district Fees

Figure 6.1 An Action Plan

b. Identified and engaged a volunteer to run the program
c. Planned the program
d. Obtained relevant supplies and materials
e Prepared and delivered promotional materials throughout the school, local youth clubs, sports clubs, and churches
f. Ensured an adequate supply of furnishings

A similar set of outcome statements would be prepared for the other elements of the after-school program.

Action planning can be an energizing process for those involved. In my recent work in East Timor I was heartened by the sense of excitement and anticipation that resulted from the development of action plans. In a Parent Teacher Association (PTA) workshop in one school, I watched the interest building as parents worked in small groups to identify activities in which they could engage to assist in improving the work of the local school. Their interest grew to excitement at the planning session, their faces glowing with anticipation as the plans emerged, so that they could clearly see the step-by-step processes through which they could accomplish their goals.

A group of tough-looking farmers, having decided they would build a small "fish pond (farm)" to raise money for the school, grew quite animated as they formulated their action plan with the assistance of a facilitator:

"What do you want to do?"

"We want to build a fish farm."

"What will you need to do first?"

"We need to find some land to build the fish farm."

"How can you get land?"

"I know someone who will donate the land."

"What will you do next?"

"We must build the dikes to hold the water."

"Who will do that?"

"We can do that ourselves."

"What else will you need to do?"

"We need a net to cover the pond so the birds won't get the fish."

. . . and so on.

After the facilitators had recorded all the information, they drew an action planning framework on a chart taped to the chalkboard and inserted details. Missing elements led to more questions that resulted in a complete plan for the development of the fish pond. The sense of pride and accomplishment among men in the group was evident, for here they could see clearly how to put their ideas into practice—a great sense of empowerment.

I have a photograph in my collection that shows the men standing by their pond, proudly showing visitors the completed project. They are clearly invested in the school, and I've heard that they have been able to sell fish in the local market to provide funds for materials needed to assist in the reconstruction of the classrooms that had so cruelly been burned by the Indonesian military.

Quality Check

The heart of community-based action research is not the techniques and procedures that guide action but the sense of unity that holds people to a collective vision of their world and inspires them to work together for the common good. The planning processes detailed earlier provide a clear set of tasks and activities, but they are not complete until these activities are checked against a set of principles. The essence of this part of the planning process is not only to check that the tasks have been described adequately but to ensure that each of the participants is aware of the need to perform them in ways that are consonant with community-based processes.

Each participant should have the opportunity to discuss his or her tasks and activities, so that all describe what they will do and the way in which they will go about doing it. Facilitators should assist this process by having participants check their activities against the criteria for the well-being of the people (see Chapter 2). Each person should be sure that his or her tasks and activities are enacted in ways that will enhance other people's feelings of

- *Pride:* feelings of self-worth (Will these activities enhance people's images of themselves?)
- *Dignity:* feelings of autonomy, independence, and competence (Are we doing things for people instead of enabling them to do them, either by themselves or with our assistance?)
- *Identity:* affirmation of individuals' social identities (Are the right people performing the tasks? Are women, for instance, performing tasks related to women's issues?)

- *Control:* feelings of control over resources, decisions, actions, events, and activities (Can people perform the tasks in their own way?)
- *Responsibility:* people's accounting for their own actions (Are we trusting them to perform the tasks in ways that make sense to them?)
- *Unity:* the solidarity of groups of which people are members (Do they have to alienate themselves from their group to perform the tasks?)
- *Place:* places where people feel at ease (Can they do the tasks in their own places?)
- *Location:* locales to which people have historical, cultural, or social ties (Do we need to relocate some of our activities?)

Facilitators also need to have people check whether their activities are consonant with the set of principles outlined in Chapter 2, which focuses on

- *Relationships:* Are processes encouraging productive working relationships between people?
- *Communication:* Are regular times for sharing information included in plans?
- *Participation:* Do major stakeholders engage in tasks set in the plans?
- *Inclusion:* Are all major stakeholders included in the plan?

The time and energy spent on this quality check constitutes a sound investment of personal resources, as they are likely to minimize delays and blockages by ensuring that all stakeholders are acknowledged and respected. The participatory and inclusive relationships enacted in action research provide the benefit of a harmonious, supportive, and energizing environment that is not only personally rewarding but also practically productive.

IMPLEMENTING

Collaborative processes often start with a flourish. Much enthusiasm and energy are generated as plans are articulated and people set off to perform their designated tasks. The best of intentions, however, often run up against the cold, hard realities of daily life. Participants in the research process reenter family, work, and community contexts, where responsibilities and crises crowd out new activities. As participants attempt to implement the tasks that have been set, research facilitators should (a) provide the emotional and organizational support they need to keep them on track and to maintain their energy, (b) model sound community-based processes, and (c) link the participants to a supportive network.

Supporting

In the early stages of activity, people often find themselves, to some extent, out on a limb. If the preliminary work has included all people affected by the research process—the stakeholders—then the performance of new tasks or different approaches to routine tasks will usually proceed with few impediments. Even so, people usually take risks when they change their usual routines and processes, and sometimes they experience criticism and disapproval. In addition, the best of plans cannot take all contingencies into account, and performing new tasks may turn out to be much more difficult than people had anticipated. To a greater or lesser extent, they will experience feelings of doubt, threat, and/or anxiety that impede their ability to continue with research activities.

In these circumstances, the press of continual work and life demands can easily lead to research tasks being delayed or neglected. The main job of the research facilitator will be to provide the practical support that will enable people to continue their research activities. Support can be provided in many ways, some of which are described briefly in the following paragraphs.

Communication

Facilitators should communicate with each participant regularly and organize simple ways in which participants with similar or related tasks can communicate. Such interactions may take the form of prearranged regular visits, telephone calls, arrangements for emergency contact, and informal social contacts—meeting for lunch, a beer after work, and so on. It is important that each person be linked with others so that participants can discuss their problems, celebrate accomplishments, maintain focus, and sustain their sense of identity with the research project.

I once assisted a group of community development workers in forming a support network for themselves after they had attended a planning conference. Because they were located in communities many hundreds of miles apart, opportunities for interaction were limited. I arranged a regular telephone conference call in which they could discuss issues, share ideas, and generally "unload." Later, each of them commented on how important that contact was. It lessened their feelings of isolation and enabled them to stay on track when other agendas and problems threatened to divert their energies.

I have also encouraged other groups to develop small support networks that can meet easily on a regular basis. Meeting for coffee, during lunch breaks, or for dinners, these groups can provide a relaxing and safe environment in which the members can talk through their problems and gain emotional support as they do so.

Personal Nurturing

As people work through their assigned tasks, their worlds are often changed in some fundamental way. They sometimes gain a deeper understanding of themselves, their need to trust both their own judgment and that of those around them, and the extent of the risks they are taking. In such circumstances, people often need personal reassurance and affirmations of their competence and worth. Facilitators should be constantly sensitive to the need to provide affirming comments to people engaged in research activities, not in a patronizing or mechanical way but authentically and specifically.

I recently asked a young teacher to speak to a university class about the work she was doing in her classroom. At first she hesitated, recommending that instead I ask another person who is an "expert" in the field. "Amy," I said, "I've heard you talk about your work and really like the ideas you have and the way you link them to the practical, daily realities of teaching. I think you would be much better for the job than an 'expert' who may have many good ideas but can't make the direct links you can. In addition, the young women in the class will be able to identify directly with you, because you are like them in many ways." Not only did she agree to speak to the class, but she also provided an excellent presentation that was applauded by the class.

Reflection and Analysis

Visits and conversations provide researchers with opportunities to ask questions that can help participants who are performing tasks describe and reflect on their activities. Such questions may touch on some of the following areas:

- *Relationships:* How have people responded to this (new) activity? Are they supportive? Has anyone caused any problems for you?

- *Patterns of work and organization:* Can you combine your activities with your work? Does this cause any problems? Do your tasks conflict with other people's ways of working?
- *Communication:* Who have you talked with about your tasks/activities? Have you talked with your supervisor/manager/administrator? Your fellow workers/teachers/clients? What have their responses been? Who would be useful to talk with from time to time? Who can support you?
- *Difficulties and solutions:* Are you having any problems? Have you overcome them?
- *Progress:* How are things working out for you? Have you made much progress? What have people been saying about your activities?

It is important that facilitators not judge the performance of the participants, even if asked to do so. There is a great difference between saying, "Things don't appear to be going very well," and asking, "How are things going for you?" Facilitators should encourage participants to review each aspect of their tasks, talking through the processes in which they are engaged and touching base on the principles, for example, "Who have you talked with about this? Will you be able to continue with this task? How are people responding to your activities?"

Assistance

When participants experience difficulties, research facilitators may need to provide assistance. They can assist directly with some activities, providing or seeking out information, doing small tasks, or acquiring needed materials. It is important that research facilitators not take over the tasks but merely provide sufficient help to enable the participant to initiate or complete them successfully. Researchers need to develop the facility to do things *with* people and not *for* them; they need to be especially wary of the temptations that arise when working with others in areas in which the researchers have expertise. It is usually more important that the people involved develop the skills to maintain the process than just that the job get done. When facilitators take over a job, they implicitly highlight the incompetence of other participants and disempower them in the process.

Conflict Resolution

Conflicts, whether minor disagreements or major arguments, are not uncommon in action research. The researcher who has maintained a relatively neutral stance in the research process can take the position of a disinterested party in a dispute. In these situations, the researcher's mediating role is to assist the parties in conflict in coming to a resolution that is satisfactory to everyone. The task is to manage the conflict so that all parties can describe their situations clearly, analyze the sources of conflict, and work toward a resolution that enables them to maintain positive working relationships.

Modeling

The ways in which research facilitators enact their supportive role will provide direct cues to other participants regarding their own ways of working. Researchers' availability and the manner in which they provide assistance and support should implicitly demonstrate community-based processes. Their openness and authenticity should illustrate the difference between a community-based approach and a patriarchal, bureaucratic, controlling style of operation.

Mahatma Gandhi said, "Be the change you wish to see in others!"

As they work with other participants, research facilitators should ensure that their procedures and working styles enact the processes and principles of community-based research. Their conversations can describe their own and other people's activities, and they should find opportunities to give "gifts"—news, information, snacks, a telephone number, or a flower. As they describe their own activities, research facilitators provide information in an analytic form that demonstrates ways in which people may reflect on their own work. For example,

> I've managed to help Mary find someone who can assist her with her project. They've already been able to . . . She says that she's really starting to feel good about that and intends to . . . She feels more comfortable now that she's able to talk with the other people in her office. Next time I see her, I'll . . .

Extended discussions provide opportunities for facilitators to pass on information but also to encourage people to reflect on their own activities. Modeling is one of the most powerful means of instituting the social processes that are inherent in community-based research. The doing is worth much more than the saying.

I recently taught a successful graduate course called Community-Based Ethnography. Of all the student feedback I received, the most consistent comments related to the way in which I taught the course. One student noted, "The instructor not only teaches about community-based research; he does it."

A member of a community in which I worked once criticized a consultant who had failed to live up to his expectations. "He can talk the talk," he said, "but he can't walk the walk."

Linking

A support network is a key ingredient in the success of a project. This is true not only for the research process itself but for each of the participants involved. As people plan their tasks and activities, they can nominate the people who are likely to support them and take steps to establish relationships with them. Participants will do much of this work themselves, but the facilitator's knowledge of the broader context will often enable him or her to link workers with other people who are sympathetic to their activities or who can provide important information or other resources. Linking participants to a supportive network provides them not only with emotional support but often with organizational and community support.

Linking participants in networks of support sometimes enables them to engage new people in research activities and extends the breadth and power of the research process itself. When people display interest, it may be appropriate to ask them if they would like to participate in activities or to help the research workers perform their tasks. In this way, linking not only extends the support network for individual participants but generates the energy that sustains a community-building process.

As research facilitators assist other participants in developing supportive links, they should be wary of inserting themselves as permanent intermediaries in the linking process. When they continue to act as "middlemen," research facilitators inhibit the development of positive working relationships between participants and others with whom they work. They maintain control and increase their own power in the situation at the expense of those they are assisting. Figure 6.2 depicts a situation in which information transfer, discussion, or interaction cannot take place except through the research facilitator. A linking, supportive network, conversely, provides multiple opportunities for exchange, conversation, and consultation (see Figure 6.3).

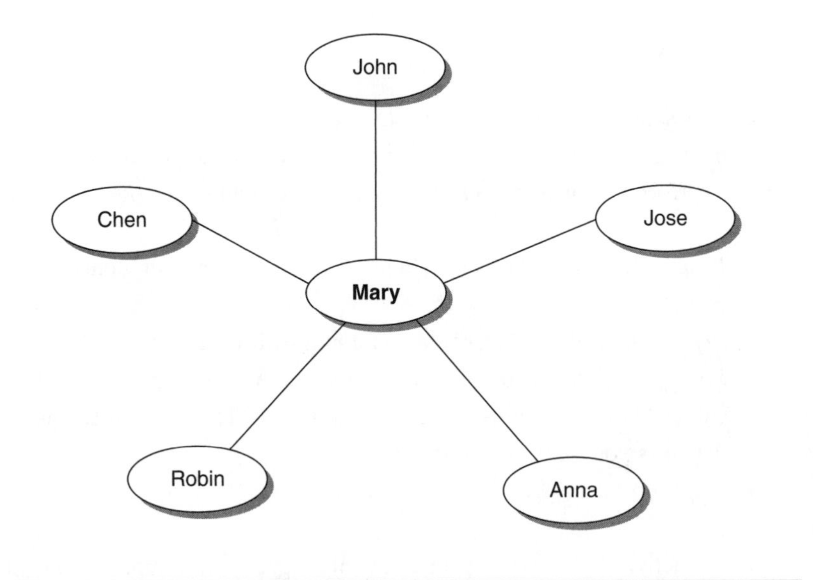

Figure 6.2 A Controlled Network

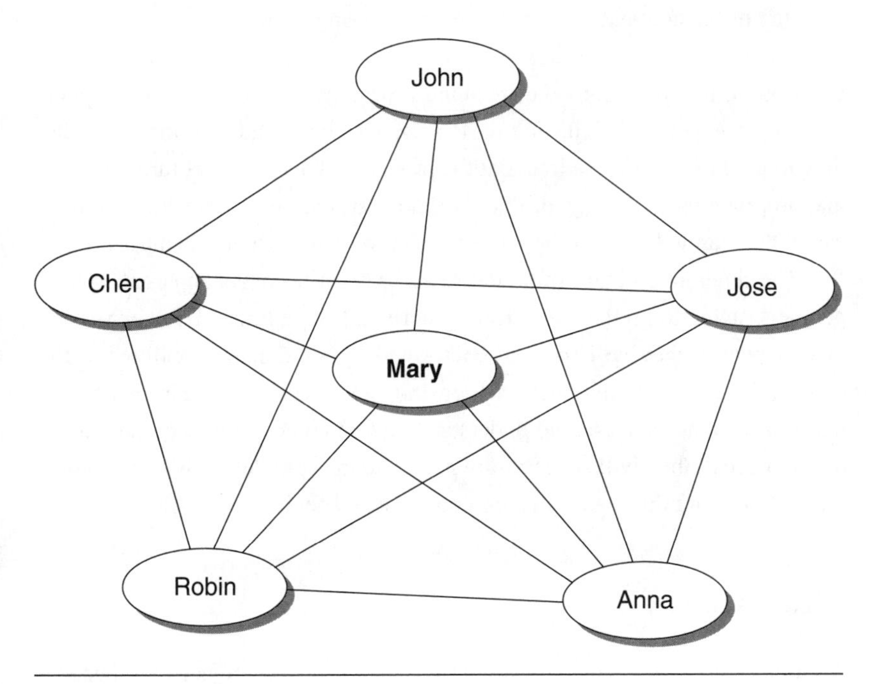

Figure 6.3 A Linking Network

REVIEWING

Participants should meet on a regular basis to review their progress. The plan should be displayed at these meetings, and stakeholders and each of the participants performing tasks should be given the opportunity to do the following:

- *Review* the plan. (Focus question: Have you had any thoughts on our plan?)
- *Report* on progress. (How are you doing with your tasks?)
- *Modify* sections of the plan, if necessary. (Are you having any difficulties? Do we need to change our approach? Do we need to change the tasks you've been assigned?)
- *Celebrate* successes. (What have we achieved?)

These activities motivate people by highlighting accomplishments or reassuring them if they have failed to make significant progress; in addition, they provide a context that reinforces their sense of community. As people strive to perform their tasks and report on their successes and struggles, they share their worlds in a real, direct way and, in the process, extend their understanding of the contexts in which they are working.

These processes also present opportunities for rethinking aspects of the description, interpretation, and planning processes and provide extra support for people experiencing difficulty. Participants may find, for instance, that they require more information, that the tasks they have been set turn out to be inappropriate in some way, or that their activities are being blocked. In these cases, the plan will need to be modified to take these exigencies into account.

The reiterative nature of action research soon becomes apparent. Where people struggle to implement activities derived from a fixed vision or version of their world, they will soon be confronted by the dynamic realities of the context. To the extent that they can construct and reconstruct their vision, taking into account the increased understanding that comes from each reiteration of the process, they will successfully negotiate the complex web of meanings, interactions, and discourses that compose social life.

EVALUATING

At some stage, the need for a formal evaluation of the project may become evident. People who contributed funds and personal or political support will probably appreciate some statement or report that provides information about the extent to which progress has been made or desired ends have been achieved.

Evaluation requires processes that are similar in nature to those used to formulate joint stakeholder descriptions and interpretations. A full treatment of the process is provided in Guba and Lincoln's *Fourth Generation Evaluation* (1989). Evaluation is carried out as a joint construction of stakeholder groups who

- Place their claims, concerns, and issues on the table for consideration
- Review information obtained from interviews, observation, documents, and group constructions
- Resolve claims, issues, and concerns
- Prioritize unresolved items

Tasks and activities that have resulted in a satisfactory resolution are delineated, and those that are unresolved become subject to continued action. As Guba and Lincoln (1989) point out, different values held by different stakeholders will lead to disagreement about priorities for further action. The research facilitator's task is to negotiate these points of disagreement and seek ways to reformulate the issues so that participants can agree on the next steps to be taken.

This method of evaluation is consonant with the constructivist philosophy that is inherent in community-based action research. It defines outcomes in ends that are acceptable to stakeholders, rather than those whose degree of success may be measured against some set of fixed criteria.

Sometimes there is considerable pressure to provide a "definitive" evaluation, especially from people who wish to use numbers to justify expenditure or their personal involvement. Numbers, however, are illusory and usually reflect a distorted vision of the research process. Nevertheless, there are occasions when some quantitative information is useful and may be properly included in an evaluation process. We may provide numbers of students enrolled in a course, youth attending a program, mothers attending a child care clinic, and so on. Numbers by themselves are misleading, however, and often oversimplify the state of affairs. They also risk reifying—creating an illusory preeminence about—certain aspects of a project or program. People may focus on tables that quantify relatively trivial features or disregard significant features of the project. Research participants should be wary, therefore, of engaging in forms of evaluation that are contrary to the principles of community-based action research. (A number of appropriate evaluation strategies are summarized by Yolanda Wadsworth in her book *Everyday Evaluation on the Run,* 1997.)

CONCLUSION

As I wrote this chapter, I described it to friends as "the sharp end of the stick." It is the point at which action occurs—where we set out to do something about the problems that have been the driving force behind all the activity. I have described routines that suggest ways to work with stakeholders to plan tasks and activities, implement them, sustain them, and evaluate them. The end point of the process should be the resolution of the problems with which we started.

Social life is rarely as simple as that, however. We usually find that myriad issues emerge when we start to poke at a problem, which can transform the problem itself and our orientation toward it. Steps taken to solve one problem sometimes take the lid off a whole range of related issues and problems. Further, the broader the context in which a problem is held, and the greater the number of stakeholders, the greater the complexity of the task confronting researchers. Chapter 7 focuses on these more complex situations and provides guidance to researchers who wish to engage deep-seated problems within relatively diverse social, organizational, and community settings.

Generally, however, clearly articulated plans based on systematic and inclusive processes of inquiry have a high probability of achieving successful results. The investment of the time required to enact action research is amply rewarded by the significant outcomes that usually result. Where people say, "Oh, I'd like to be able to do that but it would take too much time," they usually reinscribe the problem by using standard routines or procedures that have a long history of failure. Quick-fix solutions may give the impression that "we're doing something about this," but such approaches almost always fail to deal with the underlying issues that create the problems in the first place. Only by systematically revealing the reality of people's experience of problematic events, and planning concerted action that deals with the real issues affecting people's lives, can we achieve effective solutions that improve the quality of those lives.

Reflection and Practice

For this activity you should identify an issue or problem that emerged from one of the interviews, or that was apparent in the setting you observed. Alternatively, you may focus on an issue or problem from within your own work setting. Reflect on the following questions and discuss them with friends, colleagues, or classmates.

1. How would you go about formulating a solution to the issue/problem?
2. Who would need to be involved?
3. Working with a group (if possible, the people you interviewed or observed), formulate a solution.
4. Using the framework presented in this chapter, map out an action plan to enact that solution. Make sure you identify the people who will be involved and the steps, actions, and/or activities in which they will be engaged.
5. Review the action plan with your group. Identify strengths and weaknesses of the plan. Modify the plan to strengthen it.
6. Identify a process for monitoring, supporting, and reviewing the implementation of the plan. Identify who would do these activities and when they would do them.
7. With your group, map out an action plan for evaluating the planned activities.

BOX 6.1

Act: Resolving Problems

Purpose

To plan and implement practical solutions to problems that have been the focus of research

Process

Planning

Identify priorities for action.
Develop action plans that incorporate
Goals: The actions required to resolve issues
Objectives: Specific activities required to accomplish goal(s)
Tasks: The sequence of tasks required to accomplish each activity
Persons: Those who will carry out the tasks
Time: The time frame within which each task will be completed
Resources: Materials, equipment, and funds required to complete tasks

Implementing

Implement the tasks described in the action plan.
Support people systematically by
Communicating: Inform people of each other's activities.
Nurturing: Praise people's efforts.
Reflecting: Assist participants in reflecting on problems and progress.
Assisting: Actively assist people when difficulties hinder progress.
Resolving conflicts
Model appropriate behavior.
Link: Connect people in mutually supportive networks.

Reviewing

Review progress.

Evaluating

Review the plan.
Assess the effect of activities.
Revise the plan if needed.
Prioritize unresolved issues.
Celebrate achievements.

SEVEN

STRATEGIC PLANNING FOR SUSTAINABLE CHANGE AND DEVELOPMENT

Chapter 6 provided routines for moving a simple research process from the interpretive or analytic stage to practical action within limited contexts such as classrooms, offices, and small organizations. This chapter will focus on more complicated contexts, in which practitioners engage the complex issues and deep-seated problems often contained within large and diverse organizations, such as school systems, health services, government agencies, and business corporations, and/or that cover multiple community or organizational contexts in cities, regions, and states.

The intent is to ensure that changes evolving from action research processes are systematically integrated into the life of the agency, organization, department, or institution in which the research took place. Unless research participants take systematic steps to incorporate changed procedures into the ongoing life of the organization, changes are likely to be short-lived and to have little impact. Systematic and principled action research has the potential to extend its influence to all sectors of activity that have an impact on the issue or problem originally investigated. Strategic thinking and acting enables research participants to engage in significant processes of change that improve the operation of the organization and incorporate sustainable change into its operation.

In the time following the establishment of independence in East Timor, I acted as a UNICEF consultant for the Ministry of Education, Culture, Youth, and Sports. My task was to find ways of involving parents and the community in the rebuilding and management of local schools that had been devastated at the time of Indonesian withdrawal (Stringer, 2007). Initially, I facilitated participatory workshops with parents and teachers to engage them in action research processes that identified ways parents could contribute to the reconstruction and development of their local schools. At the same time, I engaged in interviews, focus groups, meetings, and workshops with directors of education, superintendents of education, school principals, and community leaders. Input from these sources provided information that was incorporated into a well-received policy paper that became the basis for developing Parent Teacher Associations (PTAs) in East Timor (Stringer, 2002).

In the following year I assisted a team of East Timorese educators in planning and facilitating a "preliminary pilot" parent workshop in one school, then left them to implement similar workshops in six pilot schools across the nation. Evaluation workshops held some months later indicated that they had been highly successful in initiating parent participation in a broad range of activities—rebuilding classrooms and furniture, making teaching materials, raising funds through gardens and fish farms, providing the school with a water supply, organizing security, and many other activities.

Workshops were then initiated across the nation by a team of trained facilitators, who planned the implementation strategies they would employ in each district. Careful and systematic processes of preparation and planning enabled district teams to systematically integrate these developments over an extended period. Thus the participatory processes of action research, initially focused on a relatively small number of people, grew into a national program that provided the basis for much needed development of the East Timorese school system.

Much of the success of this project was due to the strategic planning and support processes that included stakeholders at all levels of the school system, from senior ministry officials to parents and community leaders in local schools. Not only were the participatory processes of investigation, planning, and implementation effective at a practical level; they were also consonant with the national imperatives of this new nation. As one team member exclaimed with great jubilation, following a highly successful workshop, "This is so *democratic*!"

Although the basic "look, think, act" framework remains the same for more complex projects, practitioners need to modify and adapt their procedures and acquire additional skills to enable them to accomplish the multiple intents of action research. This chapter therefore focuses on planning and organizational arrangements that will help participants maintain control and direction of their activities in these more demanding circumstances. The following sections describe appropriate orientations to management, planning, implementation, and evaluation and the value of celebration.

MANAGING PROCESSES OF CHANGE AND DEVELOPMENT

As an integral part of a complex social system, researchers always affect other people's lives in some way when they modify their work practices or initiate new activities. Researchers are likely to disrupt practices that have long been institutionalized and that can have impacts on people's egos, dignity, power, status, and career opportunities. In almost all situations, some people will resist changes of any sort unless the processes are carefully defined and their interests taken into account.

Significant change is also likely to connect with many agencies and organizations, so that participants may find themselves subject to pressures to develop controlling and bureaucratic styles of operation that lose their community focus and override the principles of action research. Facilitators must work with participants to ensure that they are able to maintain the autonomy and integrity of their work but avoid the style, manner, and forms of operation that typify many bureaucratic settings. They need, above all, to maintain approaches to development that engage the participatory processes that foster a sense of community among all participants. This strategic approach to developmental work requires time and resources that must be factored into scheduling and budgeting of planning.

The unit for which I once worked was asked to produce a program to provide management training for people in community agencies. We estimated that development processes would take a minimum of six months and resisted efforts to get us to contract to do the job in six weeks, a period that would have allowed no time for consultation with client groups. The contract was given to another organization, which developed a set of training programs within the six weeks allotted. After six months of ineffectual activity, however, the training programs folded. We eventually acquired funding for a twelve-month developmental process, which resulted in a program that is still operating successfully today across all regions of the state.

In another instance, we estimated that we would need six months to consult local groups and formulate a redevelopment strategy for a regional community organization. Although we were funded, after negotiation, for only four months, we found ways of completing the job satisfactorily in that time. Today, five years later, that regional organization is still operating effectively.

To integrate their activities with existing organizations and agencies, participants should constantly refer to the working principles of action research. They need to inform themselves of approaches to management that are consonant with those principles so that they can enact their organizing activities in appropriate ways. Books such as Block's *The Empowered Manager* (1990), Peck's *A World Waiting to Be Born* (1993), and Senge and colleagues' *The Fifth Discipline Fieldbook* (Senge, Kleiner, Roberts, Ross, & Smith, 1994) are examples of literature that provide such orientations.

Block (1990) suggests a way of conceptualizing the difference between what he typifies as *bureaucratic* and *entrepreneurial* organizations. Bureaucratic organizations, he suggests, are characterized by patriarchal systems that emphasize a top-down, high-control orientation to organizational activity. These types of organizations breed power-oriented people who use manipulative tactics to further their myopic self-interests, which focus on advancement, approval by organizational superiors, money, safety, and increased control, rather than on service. Block further suggests that one outcome of this style of operation is the creation of cautious, dependent people who work in ways that maintain what they have.

A sign in bold capital letters on the wall of an alternative school program proclaimed, "If you keep doing what you always do, you'll keep getting what you always get."

Entrepreneurial organizations, conversely, are characterized, according to Block (1990), by people who act out of "enlightened self-interest." Enlightened self-interest focuses on activities that have meaning, depth, and substance; that genuinely serve the interests of clients or users; that have integrity; and that have positive impacts on people's lives. Entrepreneurial organizations are based on trust and a belief in the responsibility of people. The business of such organizations is managed directly and authentically, so that people know where they stand, share information, share control, and are willing to take reasonable risks. Supervision, in these circumstances, becomes oriented toward support and consultation rather than control, and success is defined as contribution to service users, clients, or customers.

Although Block's conceptual frameworks have a business orientation, his entrepreneurial approach to management emphasizes a service-oriented definition of self-interest that is in harmony with the principles of action research. He proposes that management is not necessarily as highly controlling and exploitative as is sometimes perceived, but can be enacted in consonance with more humanistic and democratic forms of organizational life.

As research projects increase in extent and complexity, the tools and resources of management become increasingly relevant. Research facilitators and other participants need organizational and management skills to supervise the wide range of activities, constraints, forces, and pressures that impinge on their activities. They also need, however, to be consciously aware of the model of management they are enacting so that they are not drawn into procedures based on traditional, hierarchical models of authority and control that damage, rather than enhance, research processes.

STRATEGIC PLANNING

Difficult problems and complex settings often require long-term and large-scale *strategic planning.* Strategic planning encompasses carefully defined and inclusive procedures that provide participants with a clear vision of their directions and intentions. It enables stakeholders to describe

- A *vision* of their long-term aspirations
- An *operational plan* that defines the particular projects or activities that will accomplish this vision
- *Action* plans that lay out the tasks and steps required to enact each of these projects or activities

A Unifying Vision

As people work toward a collective vision that clarifies the nature of the problems that have brought them together, they gain a greater understanding of the complexities of the situation in which they are enmeshed. They also gain a more holistic understanding of the multitude of factors within which problems are embedded and realize the need to formulate increasingly sophisticated plans to resolve them.

The use of drugs by youth, for instance, may encompass a whole range of factors, including education, family, lifestyle, work, leisure activities, and the media. A community seriously intent on dealing with the issue of drug abuse will need to take all these factors into account in planning. Although community members need to focus on activities that will have immediate impacts on some aspects of the problem, they also need to do "upstream" work that goes to the problem's sources rather than dealing only superficially with its manifestations.

The story goes that a man rescued a number of people that had fallen into a river. Eventually, he tired of dragging one after the other from the river and walked upstream, where he observed a bully pushing people into the water. He struggled with the bully, who was eventually arrested and taken away. The problem of the drowning people was solved by his upstream work.

When we engage in action research, we are often placed in situations where we need to be pulling people from the water and working upstream at the same time. It is important, however, that we go to the source of the problem rather than do nothing but cope constantly with its outcomes.

As their analysis reveals the factors with which they must contend, stakeholders may be able to rationalize their activities. By planning carefully, they may find that they are able to incorporate a diversity of activities into a few broad schemes or to connect a multitude of activities in ways that increase their effectiveness. A vision of the future that encompasses many facets of their common life may start to emerge.

A vision statement should clearly define the long-term aspirations of the stakeholders. It should attempt to articulate the ultimate ends of an action research project by encompassing statements about particular goals within a broader framework of ideas. Such a statement may take the following form:

> Students, administrators, and teachers at Downerton School will work with parents and other relevant community groups to provide a high quality of education for families in the district. They will develop a curriculum and a school organization that is relevant to the lives of students and enables them to
>
> - Achieve the highest levels of academic success of which they are capable
> - Gain and maintain high levels of self-fulfillment

- Develop and enact moral and ethical standards appropriate to their family and community lives
- Live in harmony with fellow students, staff, and the community at large
- Gain the skills and knowledge that will enable them to be a viable part of the workforce

The development of a vision is an attempt to integrate the many agendas that emerge from analyses in the "think" phase of action research.

The vision statement should provide the rubric under which the concerns of all participants are incorporated—academic standards, student behavior, boring teaching, irrelevant curriculum, teen pregnancy, and so on. Although different participants—teachers, administrators, students, and parents—might have different agendas and priorities, they should all identify with and accept ownership of a vision statement within which they can recognize their own agendas and interests.

Vision statements should be publicly developed as part of the planning processes outlined in Chapter 6. The previous statement would result not solely from the activities of school administration, faculty, and/or the school board, for instance, but from an extended process including all stakeholders—students, parents, community groups, business interests, and so on.

Vision statements are not the beginning point of an action research process. They arise only after considerable work has been undertaken to define specific problems and result from efforts to rationalize a variety of issues and concerns that have emanated from initial research processes. The "big plan" is an emergent reality rather than a predefined and predefining one.

Operational Statements: Enacting the Vision

An operational statement delineates the specific projects that enable participants to realize their vision. The previous vision statement, for instance, may be operationalized as follows:

The Downerton School will enact its vision through

- A school curriculum development process
- Site-based management processes
- Parent participation projects, including classroom volunteers, fund-raising, after-school programs, and short-term specific-needs projects
- A student governance organization
- A peer counseling program

- Community outreach and education programs
- Staff development programs

The intended actions stated here are based on what the various primary participants view as necessary to help them deal with their issues and concerns. Students cannot demand that their teachers engage in staff development programs; teachers should not initiate peer counseling programs for the students on their own initiative. Actions must derive from the people who are the targets of any suggested action. The teachers themselves are likely to be unresponsive to staff development programs that have been mandated without consultation, and students are likely to see a peer counseling program as yet another imposition of the adult world unless it results from their own analysis of their needs.

I once observed a group of practitioners planning a program to decrease teenage alcohol consumption in one community. Their basic premise was, "How can we stop them from drinking?" The only role they envisaged for teenagers involved using some "respectable" teens to act as models. There was no thought of involving teenagers who actually consumed alcohol. I do not know the actual outcome of the program, but I would be very surprised if the group's efforts resulted in a decrease in alcohol consumption.

There is sometimes a tendency to "gang up" on a problem by eliciting the support of many individuals and agencies. Plans to solve youth problems in a community sometimes start with meetings of concerned citizens or community leaders, who then seek the aid of social workers, teachers, other community leaders, service organizations, churches, government agencies, and politicians to develop appropriate plans. Although any or all of these may be appropriate at some stage, action research emphasizes the primacy of the principal stakeholders—those whose interests are centrally at stake. Action essentially must derive, in this case, from the youth themselves, who must ultimately formulate plans and decide what and who will be involved in the solutions they define.

Operational statements should be comprehensive and should describe the activities required to enable the primary stakeholders to accomplish the aspects of the vision that are significant to them. Participants may not be able or willing to institute all the activities at once, but the operational statement should clearly articulate all the factors that need to be taken into account to resolve the problem effectively. Research facilitators should arrange meetings that enable participants to review their vision and operational statements from time to time so that they are aware of the extent to which they have made progress toward solution of their problems. An operational statement diminishes the possibility that people will look to one-shot jobs as instant solutions that focus on one related element.

Action Plans

A separate plan should be developed for each of the activities or projects delineated in the operational statement. As each plan becomes activated, participants should define the following:

- The *objectives* of the project
- The *tasks* to be done
- The *steps* to be taken for each task
- The *people* involved
- The *places* where activity will occur
- The *timelines* and durations of activities
- The *resources* required

Processes for developing action plans are described more fully in Chapter 6.

As different teams or committees formulate their action plans, it is important that they come together to reveal their activities and directions to each other. This provides opportunities for the various smaller groups to rationalize and coordinate activities, so that plans do not work at cross-purposes or waste resources by inadvertently engaging in activities in the same areas.

Reviewing the Plans

As participants prepare to implement activities, they should appraise the strength of each plan according to the internal and external forces that impinge on it. A simple framework involving an analysis of the internal strengths and weaknesses of the plan, and external opportunities and threats, guides this process.

Strengths and Weaknesses, Opportunities and Threats

- *Strengths:* What are the strengths of the group (e.g., people and purposes)? What are the strengths of the plan (e.g., processes, resources, funds, materials, and places)?

• *Weaknesses:* What are the weaknesses of the group? Of the plan? Who is not included? What resources are unavailable? What skills or knowledge do we need to obtain? Are there any gaps in our planning?

• *Opportunities:* What can we do that we have not yet planned to do? Have we taken full advantage of the people and resources we have? Who might potentially assist us? What resources are available that we are not yet using?

• *Threats:* Who might resist our efforts? What might they do? Are we invading other people's territory? Who might perceive us as a threat?

As answers to these and other relevant questions emerge, they should be restated in the form of objectives or tasks and assigned to individuals or groups as part of their responsibilities. For instance, a weakness identified as "inadequate funds" should be restated as an objective: "to investigate possible sources of funding" or "to commence the following fund-raising activities." People can be assigned these tasks as part of the project.

Political Dimensions

In the 1960s, change processes were often driven by campaigns in which groups achieved their objectives by engaging in overtly social and political action. Community-based action research is not oriented toward this social action approach. Its purposes and objectives are to formulate links with and among parties who might be seen to be in conflict and to negotiate settlements of interest that allow all stakeholders to enhance their work, community, and/or personal life. To the extent that research facilitators are able to do this, they will increase the potential for the common unity that is at the heart of this approach to research. When researchers engage in political processes based on polarities of interest, they are likely to engage in conflictual interactions that generate antagonism. Although the potential for short-term gains is enticing, long-term enmities, in my experience, have a habit of coming back to bite you. The first impulse in action research must be to build links and formulate complementary coalitions rather than to divide the social setting into friends and enemies.

Through my work with Aboriginal people, I have become increasingly sensitive to the tendency of some individuals to constantly cast Aboriginals as oppressed peoples who are victims of colonization. To define Aboriginal life in such terms is to build a vision that has the potential to demean and diminish a group whose cultural strengths and spiritual wisdom have significance that goes far beyond their small numbers. To portray them in such terms is, from my perspective, an act of envictimization.

This does not mean that I wish to deny the exceedingly violent history that has been visited on Aboriginal people in Australia, or to diminish the social, cultural, and political problems they face. There are times when we need to confront those issues directly and forcefully. When I consider the strength and integrity of my Aboriginal colleagues and friends and the vitality of their family lives, however, I see a broader reality that goes far beyond this vision of Aboriginal-as-victim. I don't want to diminish their human potential by constantly highlighting their oppression or portraying them as victims. The words that my Aboriginal colleagues use to speak of their experience provide a vision of familial, cultural, and spiritual strength that is a much more powerful basis for action. I would rather build from that strength than struggle through a perspective based on weakness.

There may be times when obdurate, inflexible, ambitious, or fearful people will try to block the progress of a project. Research facilitators need to be aware of the political dimensions of the settings in which they work to deal with these situations. They can compile a list of those individuals and groups who are likely to assist them or to be in favor of their activities, as well as those who are likely to resist because they believe the researchers' activities to be against their interests. All these people should be included as stakeholders from the beginning of the research process.

Only when researchers have failed, despite their best attempts, to engage people in their projects, and when people have purposively set themselves against the researchers' aims, should researchers fall back on the strategies envisaged in such books as Coover, Deacon, Esser, and Moore's *Resource Manual for a Living Revolution* (1985) or engage in the "campaigning" mode suggested by Alinsky (1971) and Kelly and Sewell in *With Head, Heart, and Hand* (1988).

Financial Planning

Funding sometimes can be a contentious and awkward aspect of a project, and research facilitators need to handle finances carefully and openly to maintain the integrity of the research process. Facilitators should assist participants in formulating a clearly defined budget that links financial requirements to each of the tasks and activities defined in the original planning process. The budget must be set out clearly so that all stakeholders can understand it. They should be presented with a copy of the budget, prior to meetings if possible, and have the opportunity to discuss it during meetings.

The budget should estimate the costs involved in establishing a project as well as continuing (recurrent) costs. These may be set out as shown in Table 7.1, which is based on a budget table used in a youth leisure project.

Table 7.1 Sample Budget Table for Youth Leisure Project

	Task 1	*Task 2*	*Task 3*	*Task 4*	*Task 5*
Salaries and consultants					
Materials and equipment					
Travel					
Telephone					
Total					

An extended project that involves complex processes may require the services of someone with budgeting expertise, but participants can learn to formulate budgets for less complex projects. It is important that all costs be itemized and that decisions be made about expenditures—whether, for instance, people are to be hired for some tasks, what equipment is required, and what travel is necessary.

Finances are often the most contentious part of a process because of most people's experiences with bureaucratic organizational settings, where power and authority are invested in those delegated to control the finances. When decisions must be made about funding priorities, they should be made at meetings of stakeholders. In working through these controversial issues and the negotiation of sometimes conflicting demands, a group can establish feelings of purpose and unity. When people in positions of power make forced decisions, divisions and antagonism often result.

My high school history teacher constantly emphasized, "He who holds the purse strings holds the power!" Research participants should keep this in mind when pursuing empowering, community-based processes.

Decisions must also be made about where funds are to be held. Public finances are usually held within some incorporated body (e.g., the Gleneagle Youth Association) that has the facility to dispense and account for monies. The problem here is that the incorporated body itself has restrictions placed on the ways in which monies can be spent, with implications for the project. As a project increases in size, the group responsible for initiating activities may need to be formally incorporated to acquire and disburse funds. These are contingencies that project facilitators should take into account as they lead people through a research process.

People sometimes are tempted to develop projects or programs because grant monies are available. They either launch into projects that are only peripherally relevant to their purposes or find themselves striving valiantly to fulfill the conflicting demands of the grant and the research project. In the process, the focus of their activities moves away from the issues and concerns that provide the energy for the formation of community, and participants risk burnout, disillusionment, or loss of direction.

Participants in action research projects should formulate specific plans to ensure that they have adequate funding for the period of the project. In many instances, people set up programs, services, or facilities with seed grants (grants that are given one time only, for specific purposes) only to find that they have no continuing funding. However, people should not restrict themselves to action for which funding is available. Research facilitators need to do a delicate dance to enable participants to actualize their dreams through their own efforts without leading them into situations of failure by focusing on priorities that come from funding agencies rather than from their own analyses.

A good friend of mine used to view government funding with suspicion. "Stay away from government funding," he would say. "You'll kill yourself. Do what you can with your own resources."

Certainly, the situation of the Aboriginal community school I described in Chapter 1 is testimony to this counsel. Most of the initial work for the school came from the community's own funds, and only later did they acquire funds from other sources. The teacher who assisted the community in its formation admitted to me in conversation that she doubts whether the school would have been so successful if the people hadn't had to struggle, with meager resources, to get it going.

GUIDING THE RESEARCH PROCESS

Principles in Operation

The literature of management is replete with concepts such as *supervision, leadership,* and *control,* which are antithetical to the principles of community-based research. Hierarchical styles of management embody processes that engage centralized decision making, in which people in superior positions define courses of action to be taken, and relationships of dominance and subordination, in which obedience to authority is assumed to provide the best way to get things done.

Management of a community-based process, however, requires a different approach to operation and organization. As facilitators assist participants in organizing and implementing activities, they should consciously enact the key concepts and principles of action research, constantly providing participants with information about what is happening, maintaining positive working relationships, and including all stakeholders as active participants in planning and decision-making activities.

All participants should be aware of meeting times, places, and agendas, and facilitators should clearly articulate procedures for dealing with people's issues or concerns. Coordinators should have clearly defined procedures for passing on the outcomes of coordination meetings to their team members, and all stakeholders should be regularly informed about activities and events. All participants should be able to take advantage of opportunities to engage other stakeholders in activities and events related to the processes of inquiry.

Appropriate Language

When talking with others, people provide not only direct information but also many subtle, powerful messages that signify an orientation to the listener. The way we speak and the words we use carry implicit messages about the relative status and worth of the speaker and listener and the nature of their interaction. In general terms, therefore, it is better for research facilitators to use inclusive forms of language that have a connotation of togetherness—the first-person plural *we* rather than the first-person singular *I* or the third-person *you* or *they.* Facilitators who ask, "How do you want to do this?" subtly alienate themselves from the process. "How can we do this?" is a more inclusive form of the same question that implies that the facilitator is an equal member of the team.

For the same reason, it is important to designate different tasks using language that signals roles rather than status. Use terms such as *teams* or *groups* rather than *departments* or *divisions* (note the images those words evoke). It is better to have "convenors" or "coordinators" than "directors," "supervisors," or "heads." Both researchers' behavior and the symbolic universe within which it is encompassed should enhance the empowering intent of the research activities. The language we use has an impact on the way we work.

Examine the following questions and statements for the implicit messages embedded in them:

- How can we make kids stop wearing that sort of clothing?
- How can we get parents to become more involved in the school?
- I think that you need to develop an antismoking campaign.
- I don't like the way you've done this.

Each of these implies relationships of control, authority, and exclusion that are all the more powerful because we usually accept the form of the message as we concentrate on its content. Language orients us to our world and to each other; appropriate language thus becomes a fundamental cornerstone of action research.

Making Decisions

Significant decisions about policies, purposes, objectives, tasks, organizational structure, general procedures, and the allocation of funds and resources should be made at meetings attended by all stakeholders or their spokespersons. All stakeholders should know what is happening and have the opportunity to contribute to discussions about issues. This process sometimes can be tedious as stakeholders are brought up to speed on the organizational limitations within which many of their activities are bounded. Equality of worth, however, is bound up with equality of knowledge. If parents, clients, and other community members do not understand the circumstances under which they are working, they are not in a position to become fully active participants in the research process.

This approach may seem a time-consuming and sometimes inefficient way to go about decision making, but the investment can pay off bountifully in the long run. On the one hand, people who are clearly informed about purposes and procedures are more likely to invest themselves energetically in activities and to work tenaciously to maintain their ownership of the research process. Decisions made by an inner circle of experts, administrators, or "loaded" committees, on the other hand, are apt to be viewed with suspicion and to achieve only low levels of support.

I once attended a meeting at which three days of intensive work were required to articulate a plan for a regional organization that had become so inefficient it could no longer operate effectively. Representatives from the different stakeholding groups in the region attended this meeting and worked through the issues meticulously and deliberately.

Because extensive preparatory work had been conducted with all stakeholding groups, there was considerable agreement on most issues. Nevertheless, many details had to be worked through carefully to ensure that the functions and operations of the regional body did not interfere with the work of local groups. At each stage of the proceedings, time was allocated to allow participants, as individuals or as representatives of their groups, to voice concerns or bring forward issues that needed to be addressed.

After three days, consensus was reached about the basic constitution of the regional body, and representatives elected interim officeholders who would initiate the new body. Years later, that organization is still operating effectively, although it deals with contentious matters among groups of people who are often in conflict. The time taken to work through the issues with all stakeholders has been repaid many times over.

Research facilitators and other participants should make decisions by consensus, following procedures that ensure that everyone is clear on what is planned and how it will work. Resolutions made on the basis of a vote often leave the agendas of marginal groups neglected and sap the energy of a community. Although major decisions about policy, purpose, and objectives are made corporately by stakeholders, detailed planning of tasks and activities should be left to the people responsible for carrying them out. This provides a sense of responsibility, competence, and autonomy that heightens people's feelings of ownership of the research process.

Support and Monitoring

Chapter 6 emphasized the need to provide participants with support as they enact their action plans. This is an important function for facilitators or team coordinators, whose task is to assist participants in marshaling their energies, monitoring their activities, and maintaining a focus on their purposes. Facilitators and team coordinators should talk regularly with project members, giving them opportunities to reflect on their activities and provide feedback. Their role is not to supervise, evaluate, or judge participants' performance but to act as consultants, providing information, advice, or assistance to team members. The demeanor and behavior of facilitators and coordinators always should imply that they are resource or support persons, rather than bosses or organizational superiors.

EVALUATING

As stakeholders work through the recursive processes of observation, reflection, planning, and review, they are involved in a constant process of evaluation that enables them to monitor their activities and their progress. There may be times, however, when a formal review of a project or program is either required by a funding agency or perceived as necessary to the project. When people take the time to stand back from their day-to-day activities to explore and reflect on the processes in which they have been engaged and to share perceptions and interpretations, they gain greater clarity about the direction of their work and efficacy of their activities.

Evaluation needs to be clearly focused to achieve its desired purposes. If evaluation includes a mass of detail that is only peripherally relevant and fails to capture the crucial elements at the core of the project, then it may be counterproductive, directing attention to the wrong areas of activity and distorting the research process.

Steps to Evaluation

Purpose: Assessing the Worth and Effectiveness of Activities

Evaluation should, ultimately, assess the worth and effectiveness of a set of activities or a project according to its impacts on the primary stakeholders. Many evaluations focus on the activities in which project members engage but fail to provide any indication of the extent to which the process has made an impact on the lives of the people for whom the project was formulated.

I once helped a group of community workers in reviewing a program that engaged them in a diverse array of activities with many community groups. They enjoyed the work they were doing, and their efforts appeared to be appreciated by the people with whom they worked, but such was the extent of their activities that some of them were feeling stressed and overworked.

When I asked the community workers to estimate the extent to which their activities contributed to the purposes of the program—an increase in work opportunities for unemployed youth—they were silent. Although they had engaged in many activities with youth in the community, none could make any connection between the activities in which they were engaged and the specific purposes of their program. Young people, it seemed, did not have increased work opportunities as a result of their enterprise. The activities were pleasurable but unproductive in terms of the ostensive purposes of the program.

Audience: Who Will Read the Evaluation Report?

Prior to commencing an evaluation, the research facilitator should define the groups to whom the results of the evaluation will be reported. The processes and products will differ depending on the audiences. Is the evaluation being conducted for those who provide funds, those who control the organization, those who are the recipients of services, those for whom the project was initiated, those who provide the services, or any other stakeholder group? Answers to these questions will help the facilitator formulate a suitable evaluation process and present the resulting report in an appropriate form.

Procedures: How Is the Evaluation Carried Out?

Evaluation is an intrinsic part of the action research cycle. The period of evaluation is a time when researchers formally examine or review the processes in which they have been engaged—another cycle in the look, think, act process. In evaluation, the processes are as follows:

- Look: Describe all that the participants have been doing.
- Think: Reflect on what the participants have been doing. Note areas of success and any deficiencies, issues, or problems.
- Act: Judge the worth, effectiveness, appropriateness, and outcomes of those activities.

In keeping with the principles of action research, however, evaluation is not carried out by an outside evaluator to make judgments about the worth, effectiveness, success, or failure of a project. It is a process that enables those who have been engaged in the research project to learn from their own experience. In Chapter 6, I provided some detail on the constructivist approach to evaluation advocated by Guba and Lincoln (1989), an orientation to evaluation that is in harmony with the philosophical principles of action research. Another example is provided by Wadsworth's *Everyday Evaluation on the Run* (1997).

In some instances, agencies that provide funding will require a more directive form of evaluation to assess the extent to which project activities have attained their purposes or objectives. In this case, evaluation moves through a more direct cycle of activities. It requires participants to do the following:

- Define the purposes and audiences of the evaluation.

- Determine the goals of the project. What is the purpose? What is to be achieved? For whom? What are the intended outcomes and for whom (e.g., to increase the rate of employment of youth in Queenstown)?

- Set the objectives. What will be done to achieve the purposes (e.g., to initiate a youth employment program in Queenstown)?

- Describe activities related to the objectives (e.g., "The Youth Employment Program will increase employment opportunities for youth in Queenstown. . . .").

- Gather information indicating the activities in which participants have actually engaged.

- Gather information that will enable people to judge the extent to which the activities were successful in achieving the purposes of the project or, when purposes have not been achieved or have been poorly achieved, what might account for that poor achievement.

- Engage in processes that enable participants to make judgments about the effectiveness and worth of their activities.

Careful initial planning will facilitate the evaluation process because participants will have clearly defined their activities and the relationship between the activities in which they have engaged and the purposes of the project. Evaluation can sometimes highlight the lack of correspondence between purposes and activities. In the youth employment example given earlier, for instance, project workers had failed to articulate the relationship between their purposes and the activities in which they had engaged. They rationalized their activities as "providing good role models for unemployed youth," "improving the self-concepts of unemployed youth," and so on. Their activities may or may not have accomplished these objectives, but it is clear that they did little, if anything, to ensure that youth were actually employed. Sometimes theories—the explanations we give for events and phenomena—are inadequate because they do not fit the reality of people's lives. In this example, the participants may have improved their project by engaging unemployed youth in the process of defining the problem—that is, the reasons they were unemployed—and in finding ways that they could gain employment. Careful processes of evaluation can sometimes reveal the inadequacies of a research project's initial framing activities.

CELEBRATING

A good action research project often has no well-defined ending. As people explore their lifeworlds together and work toward solutions to their common problems, new realities emerge that extend the processes of inquiry. Problems merge, submerge, or become incorporated into larger projects. Still, there is usually a time when it is possible to stand back, metaphorically speaking, and recognize significant accomplishments. The time for celebrating has arrived.

Celebration is an important part of community-based work. It not only satisfies the human, emotional elements of the experience but also enhances participants' feelings of solidarity, competence, and general well-being. It is a time when the emotional energy expended in particularly difficult activity can be recharged and when any residual antagonisms developed during the project can be defused and relationships among stakeholders enhanced.

Celebrations should reflect the principles of action research as participants get to mingle, talk, and eat and drink together. Music and/or dance will assist the air of celebration, if the context allows it. Formal, sit-down dinners are usually not a good way of celebrating because they anchor people to tables, inhibit interaction, and usually are costly. Buffet lunches or potluck dinners are better, but barbecues, brunches, and other types of parties are equally appropriate ways of celebrating. Any celebration should be held at a time that will maximize the opportunity for all stakeholders, especially those who performed activities, to attend, and in a place where the members of the least powerful groups will feel comfortable.

Speeches should be kept to a minimum, as the purpose of the gathering is to allow participants to celebrate their accomplishments *together.* There should be a time, however, when "significant people" provide a ritualized and formal benediction to the project. Key people from within the process may speak to emphasize the collective accomplishment of all stakeholders and participants.

It may also be appropriate for other significant figures, such as the mayor or other politicians, church leaders, senior managers in government departments and community agencies, local sports heroes, or other important community figures, to contribute. The message from such people should be that the broader community or organization recognizes the participants' accomplishments and contributions.

Speeches that highlight the accomplishments of single individuals and the giving of awards to limited numbers of participants are anathema to community-based processes. When the efforts of a few individuals are recognized, those who have made "less significant" contributions may feel that their work has been inferior. In some way, all should be recognized for the contributions they have made to the success of the project. Celebration is a time when all participants can congregate to acknowledge their collective achievement and say, in one form or another, "Look what we have accomplished together."

Reflection and Practice

Most organizations, agencies, and institutions have a strategic plan, though they may not always define it as such. If you work in a school, for instance, your school may have a school improvement plan.

1. Read the strategic plan.
2. Reflect on the plan and discuss each aspect with a group of colleagues or classmates.
3. Evaluate the plan.
   - Is each section clear? Comprehensible?
   - How do you think and feel about the concepts and language within the plan?

- Can you follow the steps in the plan? Can you understand the goal(s) of the plan?
- Do you think the actions or steps in the plan are clear and achievable?

4. Using information within this chapter, describe what would need to be done by those in leadership positions to ensure that the objectives of the plan are achieved.
5. Describe other factors that might threaten the viability or effectiveness of the strategic plan.
6. Plan a process for reviewing and/or evaluating the implementation of the strategic plan.
7. How could the people in the organization celebrate their accomplishments in an authentic and effective way?

BOX 7.1

Act: Strategic Planning for Sustainable Change

Purpose

To enact procedures for organizing and managing large-scale and/or long-term change and development processes.

Process

Strategic Planning

Develop a vision statement: What is to be achieved?
Develop an operational plan: How will that vision be attained?
Develop action plans: Describe tasks and activities to be accomplished.
Review the action plans: Examine strengths, weaknesses, opportunities, and threats.
Political processes: Build cooperative links within and between stakeholding groups.
Financial plan: Construct budgets to provide for developmental and recurrent costs.

Guiding the Research Process

Leadership: Enact the principles of action research—provide information and maintain participatory, supportive relationships.
Language: Use clear language that reflects the values of community-based research.
Decision making: Engage participatory decision-making processes.
Support: Act as a resource to monitor and support people's activities.

Evaluating

Purposes: Assess the worth and effectiveness of research activities.
Audience: Decide which people will be informed of results of the evaluation.
Procedures: Describe procedures for evaluating activities.
Evaluation: Describe what has been done, what has been achieved, what is still to be done, and issues and/or agendas to be resolved.

Celebrating

Celebrate people's collective accomplishments.

EIGHT

FORMAL REPORTS

Stakeholders who operate from within institutional or organizational contexts often require formally structured reports presenting technical features of the research process and detailed descriptions and analyses of the outcomes of an investigation. Government departments, funding organizations, and universities are examples of institutions likely to require such reports. Because continuing funding of projects, programs, and services often rests on the power of reports, this chapter presents a framework for constructing formal reports that do justice to the rigor and efficacy of community-based action research.

The purpose of formal reports, theses, and dissertations is to communicate the outcomes of inquiry to major institutional stakeholding groups. They speak specifically to an academic or bureaucratic audience and must therefore be presented in forms that are acceptable within these contexts. Because of the status enjoyed by science in the modern world, many organizations, institutions, universities, schools, government departments, and business corporations try to replicate the formats required of experimental scientific research in their reporting procedures. In doing so, however, they often lose significant information or present it in a form that fails to adequately represent the complexity and significance of events or to capture the agonies, achievements, tragedies, and triumphs that constitute the reality of people's lives.

Recent developments in interpretive, action-oriented research differ in significant ways from experimental scientific investigation (see, e.g., Denzin,

1989, 1997; Guba, 1990; Lincoln & Guba, 1985) and therefore require somewhat different reporting procedures. This chapter thus suggests a framework for presenting formal reports that not only is true to the principles of community-based action research but also enables individuals, organizations, and community groups to communicate effectively the processes and outcomes of their research and development work to bureaucratic and academic audiences.

I have seen many cases in which community groups engaged in effective developmental work constantly face withdrawal of their funding. These groups lack the skills or resources to construct reports that enable those who fund their activities to understand what they have achieved or to write proposals that clearly articulate their potentials. I have been guilty of failing to provide reports that clearly described my activities in terms meaningful to departmental officers and the guidelines within which they necessarily worked. Although I complained bitterly when funding for my activities was withdrawn, I can see in retrospect that I had failed to keep one of our principal stakeholders clearly informed about my activities.

I have also listened to the agonized accounts of students and colleagues who felt assaulted by comments such as "That's not [real] research!" from people who did not understand interpretive or action research processes and the associated procedures and types of outcome. This chapter speaks to people in these situations.

THE RESEARCH ORIENTATION: ASSUMPTIONS OF INTERPRETIVE RESEARCH

The approach to action research presented in this book is derived from interpretive research processes suggested by Denzin (1989, 1997). It is based on the assumption that knowledge inherent in people's everyday, taken-for-granted lives has as much validity and utility as knowledge linked to the concepts and theories of the academic disciplines or bureaucratic policies and procedures. The intent is to concede the limitations of expert knowledge and to acknowledge the competence, experience, understanding, and wisdom of ordinary people. Action research therefore seeks to give voice to people who have previously been silent research subjects.

Like other forms of interpretive research, action research seeks to reveal and represent people's experience, providing accounts that enable others to interpret issues and events in their daily lives. In the process, researchers provide information that enables those responsible for making policy, managing programs, and delivering services to make more informed judgments about their activities, thus increasing the possibility that their policies, programs, and services might be more appropriate and effective for the people they serve. Action research reports consequently will differ significantly from those derived from prevailing studies because of the interpretive assumptions that are implicit in the approach to inquiry:

- Studies are usually limited in context, engaging processes of inquiry that focus on a specific issue or problem in a particular context.
- Researchers seek to empower principal stakeholders by engaging them as active participants in all phases of the research project, including the planning and implementing processes. It has been described as research *of, by,* and *for* the people.
- The principal purpose of the research is to extend people's understanding of an issue by providing detailed, richly described accounts that reveal the problematic, lived experience of stakeholders and their interpretations of the issue investigated.
- Stakeholder joint accounts, derived from creative processes of negotiation, provide the basis for therapeutic action that works toward resolution of the issue or problem investigated. These processes ensure tangible outcomes of direct benefit to the principal stakeholders.
- Stakeholder perspectives are placed alongside viewpoints found within the academic and bureaucratic literature.
- The outcomes of the research make the experience and perspectives of ordinary people directly available to stakeholders—professional practitioners, policymakers, managers, and administrators—so that more appropriate and effective programs and services can be formulated.

REPORTS, THESES, AND DISSERTATIONS

One of the problems that continue to concern me is the extent to which formal reports silence the voices of those of whom they speak. For years now, I have been confounded, especially in my own work, by the inability of official reports to capture the sometimes excruciatingly damaging, damning, or disappointing features of people's lives—or the moments of triumph, small or large, that signal significant, life-changing events. The ponderous language of the reports often obscures the experiences of participants and the significance of events portrayed in the body of the text. As one report notes, the words of a government report indicating that "the sewage system in the community is inadequate" do not capture the same image as "children playing around raw sewage that flows past their houses."

Reports, theses, and dissertations, therefore, have purposes that differ from those of the traditional objective accounts sanctioned by the scientific community. Recent developments in the research literature suggest new ways of formulating written reports that more effectively represent people's experience. Denzin (1997) points to the need to formulate more evocative accounts that provide empathetic understandings of events and experiences. From this perspective, research reports may look and sound more like fictional works—novels or short stories—than the impersonal, objective accounts common in many official reports.

New ways of writing, however, often are confined by traditional formats that structure reports in ways marginally compatible with the intent of the writer. Although traditional reporting structures associated with experimental science seem incompatible with the narrative forms suggested by experimental ethnographic writing, it is still possible to find ways to fulfill the needs of institutional and bureaucratic audiences while remaining true to the intent of community-based action research.

Box 8.1 compares a framework for organizing experimental and survey research reports with a framework relevant to interpretive, action-oriented inquiry. The intent of the former is to provide concise descriptions of observed relationships between variables, often related to the testing of a null hypothesis. The latter provides the means to present narrative accounts derived from the processes of community-based action research.

The remainder of this chapter presents details of the ways of reporting interpretive, action-oriented research processes. The structure is not prescriptive but suggests one way of organizing outcomes of inquiry that is acceptable to people working in institutional and organizational contexts. Researchers should feel free, therefore, to creatively play with the ideas presented, provided that they remain true to the principles of the interpretive processes presented in this book.

Box 8.1 Ways of Organizing Experimental and Interpretive Research Reports

Experimental/Survey Research	*Interpretive/Action Research*
Section 1: Introduction The introduction identifies the problem, provides background information, presents the research question(s), and states the hypothesis to be tested.	**Section 1: Focus and Framing** This section identifies the problem or issue on which the research focuses, describes the context in which it occurs, the question requiring a solution, and the objective of the study.
Section 2: Literature Review The literature review presents a summary of published research studies that have explored the problem. It identifies outcomes, gaps, and inconsistencies in the research.	**Section 2: Preliminary Literature Review** This section summarizes published research related to the issue, identifying the outcomes, perspectives, theoretical assumptions, and gaps in the existing literature.
Section 3: Methodology The methodology section presents the research design and describes the operationalization of hypotheses, sampling procedures, instrumentation, and procedures for data collection and analysis.	**Section 3: Methodology** This section describes the processes and philosophical assumptions of action research. It provides details of research procedures, including choice of participants (sampling), data gathering, data analysis, and reporting processes. It also describes procedures for ensuring rigorous and ethical research practices.
Section 4: Results The results section presents the outcomes of the study, revealing and interpreting the results of data analysis.	**Section 4: Outcomes/Findings** This section presents detailed accounts that provide an empathetic understanding of how participants experience and interpret the issue investigated. It also describes the steps taken by participants to resolve the problem studied and the outcomes of those activities and events.
Section 5: Conclusion The conclusion discusses the theoretical and practical implications of the study, sometimes also presenting recommendations for applying the results of the study.	**Section 5: Conclusion** This section compares and contrasts the participant perspectives with those in the research literature. It also presents implications of the study for policies, professional practices, and future research.

The frameworks in this chapter were derived from my attempts to assist graduate students in formulating dissertation reports within a large research-oriented university. Because most of the faculty had been trained in experimental and survey research and had little understanding of interpretive or naturalistic inquiry, they were puzzled by the students' apparent inability to formulate studies according to the dictums of hypothesis-testing research. The frameworks we designed together enabled university faculty to understand the equivalence of the research reporting procedures and thus to "make sense" of it within their own meaning structures.

I have presented this model in a number of university contexts and received positive feedback from many people, both faculty and students, and they have indicated that it provides a simple way of formulating a research report that is acceptable to the evaluative requirements of university life.

STRUCTURE OF A REPORT

This section presents a format commonly used to present the outcomes of a formal research project. Later sections of the chapter describe formats that are more appropriate to the purposes of action research processes involving community and lay audiences. The major purpose of reporting is to ensure that stakeholders are fully informed about the processes and outcomes of research activities, and the form of the report consequently will depend on the needs of particular stakeholding groups. What is appropriate for funding authorities or agency management will be different from reports required for young people, parents, or community members. A typical report uses the following format:

1. *Introduction:* Presents the problem and purpose on which the research focuses
2. *Review of the literature:* Provides an overview of current understandings or explanations of the issue investigated
3. *Methodology:* Incorporates a rationale for the approach to research and describes the research procedures used
4. *Research outcomes or findings:* Provides a narrative account of the processes and outcomes of the research—in effect, the story of participant experiences and perspectives
5. *Conclusion:* Compares and contrasts findings in the study to those presented in the literature and discusses their implications

SECTION 1: INTRODUCTION—FOCUS AND FRAMING

This section of a report presents an overview that enables readers to understand why the study was instituted and the issue on which it is focused. It frames the study by locating it within the boundaries of a particular social context—for example, classroom, home, community, or office—and describing the type of people involved—for example, students, teachers, young people, parents, and so on. The following information is presented:

- The problem or issue on which the research focuses
- The context in which that issue or problem is played out
- The question that requires an answer or solution—that asks, in effect, "Why is it so?"
- The purpose of the research—generally, to seek an answer to the research question
- The significance of the study—why the issue is important, or why the problem needs to be resolved

SECTION 2: REVIEW OF THE LITERATURE

The purpose of this section is to describe what has been learned about the issue from previous studies reported in the research and professional literature. It presents the outcomes of studies that have investigated the issue on which the current study is focused. These studies are also subject to critical analysis to reveal the concepts, theories, and underlying assumptions on which their various claims and viewpoints are based. This process digs below the surface of the reports to reveal the implicit systems of knowledge (discourses) and cultural practices embedded in the theoretical literature and in the programs and services on which they report.

The literature review includes studies reported in literature derived from

- *The academic disciplines:* Academic texts, research reports, and academic journals
- *The professions:* Publications and journals of professional bodies, such as teachers' unions, social worker associations, public service journals, and so on
- *Government agency policies and programs:* Parliamentary records, legislation, departmental documents, policy papers, annual reports,

research reports, sectional reports, client service literature, reviews, evaluations, procedural manuals, and so on

The review and analysis of this literature sets the stage for a later process (see Section 5) in which official and academic viewpoints are compared and contrasted with research participant perspectives. The review also points to gaps or inadequacies in the literature, thus revealing the need for the current study.

SECTION 3: METHODOLOGY

Introduction

The third section of a report presents a rationale for the approach to research used in the study (philosophical assumptions) and describes in detail the people involved (sample), the context in which it takes place (site), and the procedures used to conduct the research (research methods). It informs readers why this approach to research is appropriate to the issue investigated and indicates steps taken to ensure that the study was rigorous and ethical.

Methodological Assumptions: Philosophical Rationale

Because interpretive, action-oriented approaches to inquiry have been accepted only recently as legitimate in academic and official settings, the first subsection may require a more extended treatment than is expected of experimental or survey research reports. It provides information that identifies the research paradigm and provides readers with details of the purposes, processes, and outcomes of naturalistic or interpretive research. In doing so, it is useful to cite sources that enable readers to extend their understanding of the paradigm (e.g., Denzin & Lincoln, 2005; Lincoln & Guba, 1985; Reason & Bradbury, 2007).

This subsection should implicitly answer the following questions: "Why have we used this approach to research?" and "Is it a valid and rigorous research process?" It may be helpful to clarify the nature of the research process by comparing and contrasting naturalistic or interpretive inquiry with the experimental and survey research that is often presented in academic and bureaucratic reports. It is useful to know the extent to which intended readers value and have experience with interpretive action research.

Some years ago, I presented to a government department a report that evaluated one of its programs. When I formally presented the report, the head of the department informed me rather frostily that "this is not an evaluation!" Although I thought I had negotiated details of the evaluation prior to accepting the contract, the departmental officers responsible had moved on to other positions. Their replacements did not understand the nature of the report I had written, in part because of my own failure to describe the nature of my approach to inquiry in my report. I now know that my report should have clearly articulated what readers could expect as outcomes of the research process.

Research Methods

The second subsection details the way the study was carried out. It enables readers to clearly understand how researchers went about the work of investigating the issue, including who was involved, what information was collected, and how the information was analyzed. This section would be called the research design in experimental and survey research.

Position of the Researcher

It is especially important to describe and explain the role of the research facilitator. Readers should understand how the relationship between the research facilitator and other participants helped shape the processes and outcomes of the investigation. The researcher may be described as a consultant, resource person, scribe, or coparticipant whose role is to assist people rather than control them. The research is done *with* the people, not *on* or *about* them. This section is relevant to the general interpretive purpose of representing the experience and perspective of participants and to the values inherent in community-based action research. The term *participant* is a departure from the term *subject* normally used in experimental and survey research and reflects a change in the status and role of people involved. Participants in action research actively engage in monitoring and directing the processes of inquiry.

Participants

Readers should be informed of the number and type of people who participated in the investigation (called "the sample" in quantitative studies) and the way the participants were chosen (sampling procedures) so they will clearly understand the sources of information. Researchers particularly need to identify the principal stakeholders—those most affected by the issue—and other important groups who have contributed to the study.

In a school-related project, students and parents were identified as the principal stakeholders. Other significant stakeholders included the principal of the school and other principals in the district, teachers, and the district superintendent. Information from these sources enabled research facilitators to build an understanding of how the issue affected the lives of the students and their parents and how the issue fit into the local school context.

Information (Data) Gathering Techniques

As discussed in Chapter 4, this subsection describes the type of information acquired and how it was recorded. It is sometimes useful to cite authors who have suggested those procedures—for example, "This study used interviewing/data collection/data analysis procedures suggested by Spradley (1979). Initial open-ended questions that enabled participants to describe and interpret experiences in their own terms were complemented by. . . . " The following types of detail may be included:

- *Interviews:* The type of interviewing procedures used, with whom, number and duration of interviews, during what period, and where and when interviews occurred
- *Observations:* Activities, events, or locations observed, related to which people, how, at what time, for how long (or actors, acts, activities, events, objects, places, and times)
- *Documents, media, and artifacts:* Documents, official reports, minutes, procedures, materials, policies, letters, records, and so on, that were scrutinized; films, videos, and media reports reviewed; artwork, working papers, and other objects produced by participants; samples; and so on
- *Recording:* How information was recorded—field notes, audio- and videotapes, photographs, photocopies, and so on

Analysis: Procedures for Distilling and Interpreting Information

This subsection of the report provides readers with an understanding of the ways that research participants analyzed or interpreted information. It describes details of procedures used for selecting, categorizing, and labeling information (see Chapter 5). Readers need to clearly understand how interpretive procedures (data analysis) relate to the processes and products of investigation—that is, how analytic procedures provided the material on which the report is based.

Rigor

This section provides readers with evidence that the research has been carried out rigorously, that the procedures and processes of inquiry have minimized the possibility that the investigation was superficial, biased, or insubstantial. Because traditional criteria for evaluating the rigor of experimental and survey research are inappropriate, action researchers report on the set of issues that establish the study's trustworthiness (see Chapter 3)—credibility, transferability, dependability, and confirmability.

Limitations

Researchers should indicate any limitations that arose from the pragmatic realities of investigation. It is not usually possible to include all the people who should be included, to interview them for the extended periods warranted by interpretive inquiry, to follow up on all relevant issues, and to deal with all the contingencies that arise. Human inquiry, like any other human activity, is both complex and always incomplete. We need to acknowledge the extent of that incompleteness in our written accounts of the work.

Ethical Issues

This section of the report describes steps taken by research facilitators to maintain the rights and privacy of research participants, including procedures that guard against unwarranted intrusion into their lives, maintain their privacy, and establish appropriate ownership and uses of the products of investigation. It may also describe how research processes were enacted in ways sensitive to the cultural values and protocols of research participants (see Chapter 3).

SECTION 4: RESEARCH OUTCOMES/FINDINGS

Section 4 is sometimes described as the "results" section of the report, enabling researchers to present what they have discovered in their investigation. Unlike experimental research that usually reports on observed relationships between variables, interpretive research presents narrative accounts that reveal the ways people experience the issue investigated and the context within which it is held. This section presents richly detailed, thickly described accounts that enable readers to empathetically understand the lived reality of research participants. These accounts are constructed from information collected and analyzed during the study (see Chapter 5). They should include the perspectives of people from all major stakeholding groups.

Interpretive procedures described in Chapter 5 present features and elements that provide the basis for detailed, coherent accounts that

- *Describe* events, activities, and contexts from the perspective of the participants (This may include sequential accounts of events and acts, activities, objects, times, places, purposes, and emotions relevant to the issue investigated.)
- Present the way participants *interpret* events, activities, and so on, including how things occur and explanations for those occurrences (why they occurred as they did)
- Describe the *actions* taken by participants to resolve the issue or problem

These accounts may incorporate details of aspects of the study that are described in the following sections.

Setting the Scene: Describing the Context

As with any story, readers need to understand where the story is set, the people involved, and other relevant background information. One way of organizing this section of the account is to first set the scene and then tell the stories that have emerged. Another method is to begin with individual stories, weaving in descriptions of the setting as the person's story emerges. The purpose at this stage is to present a more detailed account extending the general information provided in Section 1. Descriptions should reflect participant views of the setting, rather than the "objective" voice of the researcher, so that accounts ring with participant voices, using their concepts and their language to describe the people and the setting. Any of the following descriptive features might be included:

- *Actors:* People who are significant or relevant to the story
- *Acts and activities:* The things people do, activities in which they engage
- *Events:* Significant events or incidents that take place
- *Place:* Where those activities or events take place
- *Time:* When they occur and for how long
- *Purpose:* What people are trying to accomplish, why they do what they do
- *Emotion:* How they feel about what happens, what they do
- *Objects:* Buildings, goods, materials, clothes, cars, books, reports, and so on

In setting the scene, writers should focus on those features of the situation that appear important from the participants' perspectives. This enables readers to see the context through the eyes of the participants and to understand their perspective. In some instances, large chunks of data may be included in the text to strengthen this process.

Constructing Accounts: Telling People's Stories

Common reporting practice formulates generalized accounts by combining information acquired from a number of individuals or groups. The problem with generalized accounts is that they fail to capture the lived reality, the actual on-the-ground experience of the people whose lives they are reporting. By aggregating the data, report writers risk losing those significant events or features of experience that really make a difference in participants' lives. A method more appropriate to an action-oriented, interpretive approach to research commences with stories of individuals.

As the story emerges, use descriptive features (e.g., actors, acts, and activities; see earlier list) that seem most suited to illuminating the person's experience. Focus on those aspects of the context that are significant from the person's perspective, rather than on commentary derived from others. Narrate a story that the person would recognize as his or her experience told from his or her perspective.

Accounts should provide sufficient material to enable intended audiences to understand the experience and perspectives of key people in the primary stakeholding group. The stories that emerge in the report should also provide understandings of the ways that other significant stakeholders experience the issue investigated. A report of an investigation of school dropouts not only would focus on the perspectives and experiences of students who left school early but also should provide an understanding of the perspectives of peers, teachers, parents, school administrators, and others having a stake in the issue.

Constructing a General Account

Most reports contain a section that summarizes what has been discovered during the processes of investigation. This section provides an opportunity to understand the issue in ways that enable participants to work toward a resolution of the problem(s) on which the research has focused.

Once individual accounts have been formulated, list the key features that formed the basis for each account and review them. Categorize the features according to previously described procedures, that is, identify features or themes that are common to all or a number of the stories. Names or headings may differ, but they may refer to similar features; for example, "arguing," "fighting," and "criticizing" might be incorporated within a category labeled "conflict." Extended descriptions of this process are provided in Chapters 4 and 5.

After those features or themes are identified, organize them into headings and subheadings that provide the framework for a narrative that tells how groups of people experience the issue being investigated. Accounts will focus on experiences and perspectives that people have in common but may also refer to the different perspectives people have of events. At this time, do not explore, analyze, critique, or otherwise comment on their experience. The intent at this stage is to reveal and accept nonjudgmentally and uncritically the ways people describe and interpret their experience.

SECTION 5: CONCLUSION—DISCUSSION OF FINDINGS

Whereas the previous section presented accounts of participants' experiences and perspectives, the final section places those experiences and perspectives in a broader context. This is the "so what" section of a formal report or dissertation that enables writers to articulate newly emerging understandings of the issue and to compare and contrast them with perspectives that exist in the academic literature or bureaucratic documentation. In effect, it presents succinctly what has been discovered and explores the implications of those findings. This section does the following:

- Summarizes the outcomes of the study
- Places stakeholder viewpoints in the broader social context of the issue by comparing and contrasting their perspectives with those presented within the literature
- Explores the implications of the study for policies, programs, services, and practices related to the people and the issue investigated
- Suggests actions that may be initiated or extended, or modifications of activities and/or procedures that will improve existing programs or services
- Suggests the need for further research to enhance or extend the outcomes of the current study

Its purpose is to show clearly how stakeholder perspectives illuminate the issue investigated and to suggest changes in organizational or programmatic practices implied by the outcomes of the research. In the academic world, it may also propose ways in which existing theoretical perspectives are enhanced or challenged by the new understandings emerging from the research process.

I was involved in a project that investigated the training needs of workers in a human services agency. The research clearly revealed the types of skills and knowledge required and suggested ways that skills development training could be organized. A review of existing literature on adult education and the human resources procedures of the agency augmented this information. It was creative work, requiring careful planning that encompassed the training needs revealed in the research process within the human resource development procedures of the agency. This resulted in a powerful training program that in the long term significantly enhanced the capabilities of the agency.

GIVING VOICE: ALTERNATIVE REPORT STRUCTURES

I have previously alluded to the need to develop more effective means for communicating the outcomes of research and development work. I have suggested a variety of ways in which the knowledge, the understanding, and the more pragmatic outcomes of our research efforts might be represented, including poems, dramas, narratives, fiction, memoirs, autobiographies, autoethnographies, films, and so on. This requires us as writer-researchers to draw on and develop literary devices that are able to evoke more clearly the deeply felt and experienced realities of people's day-to-day lives.

Apart from the style and nature of the language used in reports, however, their very structure often inhibits the clear and adequate representation of people's experiences and perspectives. This has recently become clear to me as I have helped people write formal dissertations. The report structure presented earlier engages both author and reader in an extensive discussion, first of the study itself, then of the literature related to the issue, and finally of the methodology and research processes. By the time readers arrive at the heart of the report, they have waded through many pages of writing that provide essentially background information. The voices of the principal stakeholders have become muted and sometimes lost in the academic or bureaucratic issues that tend to predominate in reports of this variety.

With my colleagues, I have recently commenced to work creatively to modify the structure of formal reports so that the experience and the voice of the research participants remain clearly in the forefront of the document. The frameworks in Box 8.2 are adapted from those presented in the first sections of this chapter. Although they require some modifications and adaptations to the content of some sections, the rearrangement seems more clearly suited to an approach to research seeking to give voice to those it studies.

The objective of these report structures is to provide ways of reporting that focus on the central objective of a report—the perspective of the principal stakeholders. The intent is to ensure that the most important information derived from the processes of inquiry—the perspectives, agendas, experiences, interests, and ideas of research participants—are given greatest prominence.

Structure 1 removes the methodological section from its prominent place in the report and places it as an appendix in the rear of the report. This change removes a potentially lengthy technical discussion from the body of the report, placing it in a less intrusive position in the text, while still providing that information for audiences who may be concerned about the technical features of the research process.

When Structure 1 is used, the report needs to expand the first section a little to include basic information about the nature of the report. This will inform readers of the nature of research that has been carried out and the type of results they might expect in the report. If people read this type of report expecting to see results in the form of causes and effects depicted as variables accompanied by tables of frequencies, they may not orient their reading to the type of text presented or realize the consequence of the report.

Structure 2 takes this process a step further, removing the sometimes intrusive voices of bureaucratic and academic research reports as the frame for

Box 8.2 Alternative Research Report Structure

Report Structure 1	*Report Structure 2*
Section 1: Introduction Research purpose, focus, context, and participants; brief synopsis of the methodology; the nature and structure of the report, including a brief synopsis of each chapter	**Section 1: Introduction** Same as Structure 1
Section 2: Deconstruction/ Literature Review Description and critique of existing perspectives from academic and bureaucratic literature	**Section 2: Results/Accounts** Accounts that reveal ways stakeholding participants describe and interpret their experiences of the issue studied
Section 3: Results/Accounts Accounts reveal ways participants describe and interpret their experiences of the issue studied	**Section 3: Deconstruction/ Literature Review** Description and critique of existing perspectives from academic and bureaucratic literature
Section 4: Contextualization Contrasts and comments on the differing perspectives presented in the results section and the literature review; suggests implications for policies, programs, and professional practices	**Section 4: Contextualization** Same as Structure 1
Appendix: Methodology and Research Process Describes the research paradigm and gives details of the research processes	**Appendix: Methodology and Research Process** Same as Structure 1

research participant perspectives. Thus the outcomes of the research process are reported in Section 2, leaving discussion of the literature for the following section. This enables the literature to be reviewed in light of the perspectives emerging from the research process, thus conceding the limitations of expert knowledge and emphasizing the primary relevance of the experience and know-how of people in their everyday lives.

The commonalities and differences between participant perspectives emerging in the study and the perspectives on the issue reported in the literature become one of the focal points of Section 4. We are, in effect, exploring the implications of the outcomes of the study. Where the outcomes concur with perspectives in the literature, we are able to validate, to some degree, our findings; where research outcomes are at odds with perspectives within the literature, we can explore the implications of the situation. Some people suggest that the literature review section of the report be deleted altogether, introducing perspectives from the research literature only in the final section in counterpoint to the major findings of the research and as a point of discussion for its implications.

Reflection and Practice

1. Using the framework presented in this chapter, write a report about the people you interviewed, and/or the setting you observed.
2. Give a draft of your report to the person you interviewed or the people you observed. Ask them to comment on the report's accuracy or fidelity—whether it faithfully represents their reality or perspective.
3. Using the report as a base, prepare and present a presentation for a group of your colleagues or classmates. Use some of the creative techniques for presentation suggested in this chapter.
4. Alternatively, if you are reading this text as part of a class:

   Prepare a report about your experience of the class, using techniques and processes learned in the class.

   Using some of the creative techniques suggested in this chapter, prepare a presentation for your classmates.
5. Within your work group, reflect on and discuss the reports and presentations.
6. Summarize what you have learned from the experience.

NINE

UNDERSTANDING ACTION RESEARCH

THE PLACE OF THEORY IN ACTION RESEARCH

The place of theory in action research is far from clear and differs from its place in experimental or quantitative research. For the latter, established theory "drives" the processes of inquiry, and the hypotheses to be tested are drawn from established theory. In effect researchers would say, "Given theory 'x,' what would explain the problem we are investigating?" A suggested answer to this question—the hypothesis—would then be formulated and tested, using carefully designed research procedures. Failure to disconfirm the hypothesis gives credence to both the suggested explanation and the theory from which it is drawn.

The purpose of theory in action research, however, is somewhat different. Reason and Bradbury suggest (2001, p. 451), "Theory is used to bring more order to complex phenomena, with a goal of parsimonious description so that it is also of use to the community of inquiry." For them, a good interpretation—theory—is one that is more reasonable than others. And reasonableness can be tested in the community of inquiry. "A new theory enables us to 're-see' the world, or see through taken-for-guaranteed conceptual categories that are oppressive or no longer helpful."

Earlier versions of action research assumed that there was a direct logical connection between theory and practice, but as Gustavsen (2001) points out, there can be no question of a direct relation. While theory can inform or influence practice, he suggests, "The link is a discursive one where ideas, notions, and elements from the theory can be considered in the development of practice but with no claims to being automatically applicable" (p. 18). "Theory," he goes on to suggest, "can inform a process of enlightenment and out of this process can emerge new practices."

There is another element to this relationship between theory and practice. The explanations and theories of the academic world often do not fit comfortably into people's everyday reality. The language, forms of propositional knowledge, and sometimes arcane idiom of academic texts are frequently inaccessible to a lay audience. Academic theories are embedded in a set of concepts, assumptions, and views of reality that make sense only within a particular social context—in this case, the discourses of the academic world.

Another reason why academic theory may not "make sense" to people is that *any* theory is just one in a wide set of possibilities for explaining or interpreting events in the social world. The particular lens through which we view the world will affect what we see. In astronomy, for instance, different "lenses" collect particular forms of light, particles, or energy. A lens that is designed to collect light in the infrared spectrum will show the same celestial object quite differently from a lens that collects X-ray or radio sources of information. So it is with social theory. Different theories will focus on different aspects of the situation and interpret the information according to the assumptions and orientations of the theory.

It is not that any theory is "wrong" or "right" but that it focuses on particular aspects of the situation and interprets them in particular ways. We might look at an event and interpret it in terms of colonialism, gender domination, racial politics, power relationships, personality, socially learned behavior, prejudice, or any number of other ways of conceiving and interpreting the situation.

One of the strengths of action research is that it accepts the diverse perspectives of different stakeholders—the "theory" each will hold to explain how and why events occur as they do—and finds ways of incorporating them into mutually acceptable ways of understanding events that enable them to work toward a resolution of the problem investigated. People from outside the research context who impose their own theories without having a deep understanding of the nature of events and the dynamics of the context are likely to either misrepresent or misinterpret the situation. They are, in effect, prejudging the situation on the basis of limited understandings, even though they may feel they can see *exactly* what is happening.

One of the tasks of action research, then, is to ensure that the various ways that different stakeholders describe and interpret events become the central focus of the research process. These different perspectives consequently become subjects of interaction and negotiation as people creatively explore ways of conceiving the situation in ways that assist them in resolving a problem.

Having said that, however, it is not possible to enter a research process atheoretically. That is, each person involved, including research facilitators, will have assumptions about appropriate ways of enacting research processes that are drawn from their own perspectives and values. The approach to research presented in this book, for instance, is based on assumptions that are articulated within the first chapters as "working principles." A fundamental assumption is that the procedures and processes of research work to ensure that power differentials do not undermine the need to clearly hear the voice of less powerful groups.

The theory, in this case, is a theory of *method* and provides clarity and understanding about the way participants enact processes of inquiry in order to achieve the practical and effective outcomes we desire. This type of theorizing we must do prior to and during the research to ensure that we can clearly articulate our purposes and processes. It is quite distinct from theories about the nature of events surrounding the research issue. Though academic theories may provide interesting interpretations of events, they are likely to distort and misinterpret the situation if they take precedence over stakeholder theories in practice. Theories that are more relevant and effective emerge from the hermeneutic dialectic—meaning-making dialogues—between stakeholders, using the concepts, terminologies, and formulations that "make sense" to them.

The following commentary, therefore, focuses on methodological theory—the theories that inform and clarify the nature of the action research processes presented in this book.

THE THEORY BEHIND THE PRACTICE

Community-based research starts, as does all research, with a problem to be solved. Unlike positivistic science, however, its goal is not the production of an objective body of knowledge that can be generalized to large populations. Instead, its purpose is to build collaboratively constructed descriptions and interpretations of events that enable groups of people to formulate mutually acceptable solutions to their problems. Community-based research, however, recognizes that any research process has multiple outcomes and takes into account the need to enact ways of working that protect or enhance the dignity and identities of all people involved. It is oriented toward ways of organizing and enacting professional and community life that are democratic, equitable, liberating, and life enhancing.

The approach to inquiry that I have presented may appear somewhat idealistic. Competitive social values, impersonal work practices, and authoritarian modes of control impose themselves as constant conditions of our work. Institutionalized practices—those commonly accepted as "the way things are done"—seem so pervasive and normal that we often cannot envisage other ways of working, even when those practices are ineffective or, in some cases, detrimental to our purposes.

I once engaged in a study of classrooms by observing three primary school classes for a year. Although at that time, I had been a classroom teacher for almost 10 years, the study dramatically changed my vision of schools. My discovery that children spent extended periods passively watching or listening resulted in a significant change in my perspective. As a classroom teacher, my vision of classrooms was one in which there was a constant and sometimes almost overwhelming flow of activity. I was astounded that young children were forced to cope with long periods of inactivity during which they were permitted only to watch and listen as the teacher interacted with other individuals. That experience changed my perception of classroom life quite fundamentally, so that I am now much more sensitive to the passivity that is a pervasive feature of school life and to the need to engage students constantly in active learning processes.

The pressures of bureaucratic life, the limitations of resources, and the competitive push for career advancement consistently work against attempts to humanize and democratize research processes. Community-based procedures that highlight the active participation of people in formulating and controlling research activities and events are hard to attain and difficult to maintain. When they can be achieved, however, they provide a powerful means of accomplishing any set of social or professional goals. The act of observing and reflecting on our own practices can be an enlightening experience, enabling us to see ourselves more clearly and to formulate ways of working that are more effective and that enhance the lives of the people with whom we work. Collaborative processes not only generate a sense of purpose and energy but also provide the means for the accomplishment of goals and the solution of problems and produce conditions that enhance participants' personal, social, and professional lives.

"BUT IT'S NOT SCIENTIFIC": THE QUESTION OF LEGITIMACY

Although action research is gaining increased support in the professional community, it has yet to be accepted by many academic researchers as a legitimate form of inquiry. From time to time, practitioners who engage in action research will find themselves subject to the negative comments, sometimes quite vitriolic, of people who do not regard such work as genuine research because "it's not scientific." A recent proposal for a paper on action research that I submitted for presentation at a national educational research conference, for instance, was rejected after one reviewer commented, "There may be a place for this nonsense in AERA [American Education Research Association], but not, hopefully, in this [the research] division."

It may be worthwhile, therefore, to spend a little time here exploring issues related to action research, so that those who enact it may rest easy in the knowledge that despite such comments and questions, action research is a legitimate, authentic, and rigorous approach to inquiry. Fuller treatments of this debate are provided in a number of other books (e.g., Reason and Bradbury's *Handbook of Action Research* [2007], Denzin and Lincoln's *Handbook of Qualitative Research* [2005], and Lincoln and Guba's *Naturalistic Inquiry* [1985]), but an overview will at least provide a basic understanding of the nature of the issues to be addressed.

I have studied the issues related to the different forms of social research for many years. In that time, I have become aware that the field is sufficiently contentious to engage the attention of some of the world's leading scholars. The debate is far from concluded, although the weight of argument seems to suggest that what has traditionally been accepted as scientific research is but one of a number of legitimate approaches to academic and professional inquiry.

Whether action research is accepted as scientific depends on the way in which *science* is defined. Certainly it is, in one sense, rigorously empirical, insofar as it requires people to define clearly and observe the phenomena under investigation. Levin and Greenwood (2001) suggest that action research is emphatically scientific, though not in terms propounded by experimental or quantitative researchers. For them, "the nucleus of scientific activity is deliberative, democratic sensemaking among professional researchers and local stakeholders" (p. 105). It is evident, however, that action research does not follow the carefully prescribed experimental procedures that have become inscribed as *scientific method.*

Experimental method, a traditional approach to scientific inquiry, seeks to test theories that purport to explain why or how the world is as it is. The ultimate aim is to derive lawlike statements that explain the nature of the world or the nature of reality. This methodology seeks to generate knowledge that is *objective* (not amenable to the subjective or authoritative judgments of individuals, organizations, or institutions) and *generalizable* (applicable to a wide variety of contexts). It must also be *reliable* (the results should be replicable by any person similarly placed) and *valid* (it should describe a *true* state of affairs). The laws of science seek to provide invariant forms of knowledge that enable us to predict future events on the basis of a preexisting set of conditions. In recent history, this ability to predict and therefore control many facets of the physical world has provided humanity with the ability to manipulate the environment to an unprecedented degree. The technological miracles of the modern world are a testament to the power of knowledge derived from the application of the scientific method.

The success of this form of scientific inquiry in the physical world, however, has not been mirrored in investigations of human behavior and the social world. Except to the extent that humans are physical beings, scientific investigation has largely failed to provide a social equivalent of the comparatively stable body of knowledge about the physical universe. A science of humanity, social life, or individual behavior has failed to emerge within anthropology, sociology, or psychology, despite the huge resources poured into research in these disciplines in the past few decades. Human beings, it seems, are hard to predict and difficult to control.

Two major factors account for the failure of the scientific method to provide a set of laws for human behavior and social life. First, scientific knowledge is now recognized to be much less stable, objective, and generalizable than previously assumed and therefore is less secure as a basis for formulating human action. Second, there is increasing acceptance of the fundamental difference between the nature of the social world and that of the physical world. Although it has been relatively easy to accept the notion of a fixed reality that could be *discovered* in the physical universe, the social universe is now recognized as a continually evolving cultural creation. Social reality exists as an unstable and dynamic construction that is fabricated, maintained, and modified by people during their interaction with each other and their environments. It operates according to systems of meaning embedded in each cultural context and can be understood only superficially without reference to those meanings. Investigation of the social and behavioral worlds cannot be operationalized in scientific terms because the phenomena to be tested lack the stability required by traditional scientific method. Humans are, to a large extent, what they define themselves to be in any given situation.

So where does this leave those who wish to investigate human affairs in order to provide knowledge that will "make a difference" to people's lives? In the past, it has been accepted that "experts," or those who are authorized by credentials that signal their access to scientific knowledge, should provide answers to social problems. Researchers are becoming increasingly aware of the limitations of this perspective, however. Scientific knowledge is partial, incomplete, and reductionist (i.e., it reduces phenomena to minute components), and while it can contribute to an understanding of elements of the social world, in and of itself it cannot provide a comprehensive explanation of events.

Professional practitioners, armed with a set of practices purportedly derived from scientific theories about human behavior, often believe their activities are based on objective knowledge. Though scientific knowledge certainly contributes useful information, the routines and recipes that compose professional practice are imbued with concepts, constructs, values, and perceptions derived from the particular social histories and cultural experiences of the individuals and groups that dominate professional and institutional arenas. They are also heavily influenced by the institutional imperatives of the organizational contexts within which practitioners work. As a result, programs and services often fail to provide for the real needs of the people, especially with marginalized or disempowered social groups. In some cases the failure to recognize the deep-seated disconnection of the service from the social reality of the people aggravates the problems the service was meant to remediate. Programs and services often deny the strengths that are an inherent part of their community and family lives, and welfare or educational services are often accompanied by degrees of implicit criticism and control that attack people's very humanity.

The strength and vitality of Aboriginal family and community lives, which I experience in a continuing way as I work in organizational and community contexts, is belied by the perspectives presented in the mass media. There is a tendency to focus on that minority of Aboriginal people who capture the sensationalizing gaze of the media, to generalize from that group to all Aboriginal people, and to fail to recognize and value the contributions of the many solid and hardworking persons who make up the majority of the Aboriginal community. Professionals tend to approach Aboriginal people with the intent of "helping" them or tend to believe them in need of "training." They talk of the "cultural deficits" of Aboriginal children and the need for Aboriginal people to learn "social skills" or to acquire "cultural capital." The ethnocentrism of this group of professionals would be merely annoying if it were not for the fact that many of them are in positions of power that permit them to have significant impacts on the lives of people for whom they have so little regard. Aboriginal people often find their lives controlled by experts and public servants who have little understanding of their social and cultural realities and are apt to act in ways that are inappropriate or demeaning from an Aboriginal perspective.

I have extended experience of similar dynamics in the United States, where Anglo or middle-class professionals fail to understand the deep social and cultural dynamics that affect the lives of their students and clients. African American, Hispanic, and Native American peoples continue to struggle against the tide of ethnocentrism that flows so strongly through the fabric of institutional life.

I am once again reminded of the words of Aboriginal social worker Lilla Watson as she spoke to a group of non-Aboriginal social workers: "If you've come to help me, you're wasting your time. But if you've come because your liberation is bound up with mine, then let us work together."

POWER, CONTROL, AND SUBORDINATION

One of the fictions of modern professional and organizational life is that scientifically based procedures will provide the means to achieve effective outcomes in any form of service—health, welfare, education, business, and so on. "Scientific" management processes purport to identify a "best practice" and formulate processes for applying it in precise detail. This is a continuing source of frustration for those whose duty is to perform these services or to gain the outcomes stipulated by government departments and agencies, educational institutions, and health services for which they work. Centrally devised best practices rarely can take into account the dynamic social and cultural forces that operate in diverse contexts in which professional practitioners work and therefore place them and their clients and students in untenable situations. These groups are subject to increasing levels of stress and disenchantment as restrictive legislative mandates and highly prescribed administrative controls attempt to dictate precisely the ways they enact their professional duties. Managers and administrators of programs and services are often subject to even greater stress, caught in the nexus of organizational imperatives, recalcitrant subordinates and clients, the complex reality of the social contexts they engage, and their personal needs for ego satisfaction and career advancement. Highly prescriptive plans provide little opportunity for managers and practitioners to adapt and adjust their work to the realities of the particular environments in which they operate and act to increase the levels of anxiety associated with their work.

The problem is, of course, that it is impossible to control human behavior with the rigor and precision demanded by the procedures of the physical sciences. The dynamism of social life and the creative and willful facets of human behavior prevent the high degrees of control that are embedded in scientific method and technological production. Attempts to impose the same type and extent of control in the delivery of human services have led to increasing levels of stress and alienation as practitioners struggle to provide necessary services within the boundaries of increasingly restrictive policies and procedural rules.

I rarely meet a friend or colleague who is "happy in the service" these days. Those who work in human service agencies or organizations as practitioners or managers frequently complain about the unrealistic demands and excessive numbers of directives that inhibit the effective accomplishment of their work. The complaints cover all fields, from social work, through education and health work, to community and youth work. The quality of professional life appears to be diminishing as demands for greater control and accountability increase.

How, then, are we to understand the events that make up our lives? Behavioral theorists, on the one hand, focus on the capabilities and characteristics of individuals and point to factors such as motivation, achievement need, intelligence, and cognition to explain people's behavior. Social theorists, on the other hand, tend to stress the large-scale forces—class, gender, race, and ethnicity—that influence social events. Marxist-oriented theorists, for instance, explain social events by the controlling and competitive relationships inherent in modern capitalist economic systems. The people who control the means of production, this genre of theory suggests, maintain systems of domination that reinforce the power and authority of those in positions of power at the expense of subordinate groups.

Although these theories are interesting and useful for some purposes, they are limited in their applicability to the problems experienced by individuals and social groups in their day-to-day lives. If we accept the economic determinism inherent in Marxist theory, for example, we are but puppets on the world stage, our lives ordered by the workings of the international economic order, our efforts to improve our lives essentially fruitless.

A recent genre of theory, known collectively as *postmodern,* provides a distinctive way of envisioning the social world that enables us to understand human experience in different ways. Although modern perspectives of the world are bound to scientific visions of a fixed and knowable world, postmodernism questions the nature of social reality and the very processes by which we can come to know about it. Elements of postmodernism suggest that knowledge can no longer be accepted as an objective set of testable truths because it is produced by processes that are inherently "captured" by features of the social world it seeks to explain. Scientists, as products of particular historical and cultural experiences, will formulate explanations of the social world that derive from their own experiences and hence tend to validate their own perceptual universes.

From a postmodern perspective, attempts to order people's lives on the basis of scientific knowledge largely constitute an exercise in power. Knowledge, and the research that produces that knowledge, is as much about politics as it is about understanding. An understanding of what research is about is not just an exploration of method but an inquiry into the ways in which knowledge is produced and the benefits that accrue to people who control the processes of knowledge production.

The theoretical discourse that follows is not presented as a description of "the" social world. It is articulated as one way of interpreting social experience and presents perspectives that seem to make sense from my viewpoint. In essence, I, as author, provide a framework of ideas that is a rationale for community-based action research. Those readers who find it unhelpful may choose to skip the next passage and formulate their own rationale for accepting or rejecting the approach to research that I advocate.

UNDERSTANDING POWER AND CONTROL: POSTMODERN PERSPECTIVES

Postmodernism derives much of its power from the way it deconstructs—that is, pulls apart for examination—mechanisms of knowledge production. Scientific rationality, a key feature of modern life, has itself become subject to forms of philosophical investigation that raise doubts about many of the taken-for-granted assumptions on which our understandings of the social and physical universe rest. An exploration of postmodern perspectives, therefore, assists us in extending our understanding of the frustrations experienced by practitioners as they work in social, community, and organizational contexts.

Michel Foucault's (1972) exploration of social life, for instance, reinforces the notion that there can be no objective truth, because there is an essential relationship between the ways in which knowledge is produced and the way power is exercised. Foucault's study of the development of modern institutional life led him to conclude that there is an intimate relation between the systems of knowledge—discourses—by which people arrange their lives and the techniques and practices through which social control and domination are exercised in local contexts. Humans are subject to oppression, Foucault suggests, not only because of the operation of large-scale systems of control and authority but also because of the normally accepted procedures, routines, and practices through which we enact our daily public and personal lives.

From a Foucauldian perspective, such institutional sites as schools, agency offices, hospitals, clinics, and youth centers might be viewed as examples of places where a dispersed and piecemeal organization of power is built up independent of any systematic strategy of domination. What happens in these contexts cannot be understood by focusing only on systemwide strategies of control. At each site, a professional elite, which includes administrators, researchers and teachers, social workers, nurses, doctors, and youth workers, defines the language and the discourse and, in doing so, builds a framework of meaning into the organization and operation of the system. Individual members of this elite exert control by contributing to the framing and maintenance of ordinary, commonly accepted practices, which are often enshrined in bureaucratic fiat, administrative procedure, or government regulation. The end point of this process is the accrual of "profit" or benefit to people in a position to define the "codes of knowledge" that form the basis for organizational life. Professional acceptance, employment, promotion, funding, and other forms of recognition provide a system of rewards for people able to influence or reinforce definitions inscribed in reports, regulations, rules, policies, procedures, curricula, texts, and professional literature.

Foucault (1972) contends that any large-scale analysis must be built from our understanding of the micropolitics of power at the local level. For him, any attempt to describe power at the level of the state or institution requires us to

> conduct an ascending analysis of power, starting . . . from its infinitesimal mechanisms, which each have their own history, their own trajectory, their own techniques and tactics, and then see how these mechanisms of power have been—and continue to be—invested, colonized, utilized, involuted, transformed, displaced, extended, etc., by ever more general mechanisms and by forms of global domination. (p. 159)

If we accept Foucault's analysis, then many negative features of society are intimately related to the ways in which people organize and act out their everyday lives. Feelings of alienation, stress, and oppression are as much products of everyday, taken-for-granted ways of defining reality and enacting social life as they are the products of systems that are out of people's control. The means by which people are subjugated are found in the very "codes" and "discourses" used to organize and enact their day-to-day lives. Oppressive systems of domination and control are maintained not by autocratic processes but through the unconsciously accepted routine practices people use in their families, communities, and occupations.

Foucault suggests that the only way to eliminate this fascism in our heads is to explore and build on the open qualities of human discourse and thereby intervene in the way knowledge is constituted at the particular sites where a localized power-discourse prevails. He maintains that people should cultivate and enhance planning and decision making at the local level, resisting techniques and practices that are oppressive in one way or another. Foucault (1984) instructs us to "develop action, thought and desires by proliferation, juxtaposition, and disjunction" and "to prefer what is positive and multiple, difference over uniformity, flows over unities, mobile arrangements over systems. Believe that what is productive is not sedentary but nomadic" (p. xiii). He suggests that we become more flexible in the ways we conceive and organize our activities to ensure that we incorporate diverse perspectives into our social and organizational lives.

The theme of control is also taken up by Fish (1980), who describes social life as "interpretive communities" made up of producers and consumers of particular types of knowledge or "texts." Within these communities, individuals or groups in positions of authority control what they consider to be valid knowledge. Classroom teachers, administrators, managers, social workers, health professionals, administrators, policymakers, and researchers are examples of "producers" who control the texts of social life in their professional domains. In organizing classrooms, writing curricula, defining the rules and procedures by which services operate, formulating policies, and so on, they control the boundaries within which particular interpretive communities will operate. They have the power to dominate the ways in which things happen in their particular domains.

Fish's position is complemented by the work of Lyotard (1984), who casts doubt on the possibility of defining social and organizational life according to well-ordered, logical, and objective (read "scientific") systems of knowledge. He suggests that people live at the intersection of an indeterminate number of language games that, in their entirety, do not constitute a coherent or rational order, although each game operates under a logical set of rules. His vision of a social world atomized into flexible networks of language games suggests that each of us uses a number of different games or codes depending on the context in which we are operating at any given time. There is a contradiction, Lyotard suggests, between the natural openness of social life and the rigidities with which institutions attempt to circumscribe what is and is not admissible within their boundaries.

Derrida's (1976) notion of the interweaving of discourses provides yet another perspective on the texts of social life. Derrida provides insight into why there is a continuing tension between people in positions of control and their subordinates and clients. According to him, cultural life can be viewed as a series of texts that intersect with other texts through the processes of social interaction. In portraying written texts as cultural artifacts—that is, as human productions—Derrida suggests that both reader and writer interact on the basis of all that they have previously encountered. Both author and reader participate separately in the production of meanings that are inscribed in and derived from that text, although neither can "master" the text—that is, control the meanings conveyed or received—in any ultimate sense. Writers tend to accept the authority to present reality or meaning in their own terms, but these meanings are deconstructed and reconstituted by readers according to their own experiences and interpretive frameworks.

In the context of organizational and institutional life, people such as administrators, teachers, social workers, and other practitioners who control the rules and procedures by which their subordinates, students, or clients work or live often fail to understand that each of the participants will understand those different texts according to his or her own framework of understanding. Even when participants accept what is said or written, they may enact their interpretations of the text in ways that seem to conflict with the perspectives of the authors of the text. Much of the apparent recalcitrance or wrong-headedness of workers, students, and clients might be understood as merely misinterpretation.

There is an implicit ideological position in Derrida's writing. He suggests the need to find new ways of writing texts—rules, procedures, regulations, forms of organization, reports, plans, and so on—to minimize the power of people in positions of authority to impose their perceptions and interpretations. Thus he implies the need to structure organizations in ways that create greater opportunities for popular participation and a more democratic determination of the cultural values embedded in the procedures that govern people's lives. This also implies the need to enact knowledge-producing activities in ways that are more participatory and democratic, enabling the perspectives and agendas of client groups and students to be included in the development of programs that serve them.

Huyssens (1986) also speaks to these issues. He is critical of writers whose theorizing—systems of explanation—presumes to speak for others. He suggests that all groups have the right to speak for themselves, in their own voices, and to have those voices accepted as authentic and legitimate. The authenticity of these "other worlds" and "other voices" is an essential characteristic of the pluralistic stance of many postmodern writers. Huyssens's position has much in common with those of writers such as West (1989) and Unger (1987), who place a premium on the need to educate and be educated by struggling peoples. This stance reflects a movement within the postmodern tradition that shifts the focus of scholarship away from the "search for foundations and the quest for certainty" (West, 1989, p. 213) toward more utilitarian approaches to the production of knowledge.

West (1989) provides a compelling argument for a more pragmatic approach to our ways of understanding the social world. His notion of "prophetic pragmatism" points to the need for an explicitly political mode of cultural criticism. He suggests that philosophy—more generally, intellectual activity or scholarship—should foster methods of examining ordinary and everyday events that encourage a more creative democracy through critical intelligence and social action. He advocates ways of living and working together that provide greater opportunities for people to participate in activities that affect their lives. He urges philosophers—academics, researchers, experts, and professionals—to give up their search for the foundations of truth and the quest for certainty and to shift their energies to defining the social and communal conditions by which people can communicate more effectively and cooperate in the processes of acquiring knowledge and making decisions.

The underlying notion in West's (1989) work is not that philosophy and rational deliberation are irrelevant but that they need to be applied directly to the problems of the people. West's pragmatism reconceptualizes philosophy—and therefore research—as "a form of cultural criticism that attempts to transform linguistic, social, cultural and political traditions for the purposes of increasing the scope of individual development and democratic operation" (p. 230). He advocates fundamental economic, political, cultural, and individual transformation that is guided by the ideals of accountable power, small-scale associations, and individual liberty. This transformation can be attained, he implies, only through the reconstruction of the practices and preconceptions embedded in institutional life. On the political level, West acknowledges the need for solidarity with the "wretched of the earth," so that by educating and being educated by struggling peoples we will be able to relate the life of the mind to the collective life of the community.

West's emphasis on liberation links him, conceptually, with the German scholar Jurgen Habermas, who, although often not classified as postmodern, has provided important ideas that can assist us in understanding human social life. Habermas (1979) focuses on the need to rethink the cultural milieu in which working-class people attempt to find meaning and satisfactory self-identity. He suggests that we clarify the nature of people's subjection and seek human emancipation from the threats of military conflict and dehumanized bureaucratic domination through more effective mechanisms of reflection and communication. Habermas proposes that the emphasis in institutional and organizational life on the factual, material, technical, and administrative neglects the web of intersubjective relations among people that makes possible freedom, harmony, and mutual dependence. His universal pragmatics attempt to delineate the basic conditions necessary for people to reach an understanding. The goal of "communicative action" is an interaction that terminates in "the intersubjective mutuality of reciprocal understanding, shared knowledge, mutual trust and accord with one another" (p. 3).

The general thrust of the ideas presented earlier is to question many of the basic assumptions on which modern social life is based. In general, these ideas are in opposition to rigidly defined work practices, hierarchical organizational structures, representation in place of participation, the isolation of sectors of activity based on high degrees of specialization, centralized decision making, and the production of social texts by experts or an organizational elite. Inversely, these perspectives suggest emphasis on the following:

- Popular and vernacular language and its meanings
- Pluralistic, organic strategies for development
- The coexistence and interpenetration of meaning systems
- The authenticity of "other worlds" and "other voices"
- Preference for what is multiple, for difference
- Flexibility and mobility of organizational arrangements
- Local creation of texts, techniques, and practices
- Production of knowledge through open discourses
- Flexibility in defining the work people will do
- Restructuring of relations of authority

Postmodern thought moves us, therefore, to examine the ordinary, everyday, taken-for-granted ways in which we organize and carry out our private, social, and professional activities. In the context of this book, it demands that we critically inspect the routines and recipes that have become accepted and commonplace ways of carrying out our professional, organizational, and institutional functions. By illuminating fundamental features of social life, postmodern writers provide us with an opportunity to explore social dimensions of our work and to think creatively about the possibilities for re-creating our professional lives.

The influence of postmodern perspectives now can be seen throughout the academic disciplines and in the professional literature. The editors of a recent edition of the *Handbook of Qualitative Research* (Denzin & Lincoln, 2005) suggest that "postpositivist inquirers of all perspectives and paradigms have joined in the collective struggle for a socially responsive, democratic, communitarian, moral and justice-promoting set of inquiry practices" (p. 1123). They further suggest, "The search for 'culturally sensitive' research approaches—approaches that are attuned to the specific cultural practices of various groups and that 'recognize ethnicity and position culture as central to the research process' (Tillman, 2002)—is already under way" (p. 1123). These suggestions stand alongside their advocacy of a new praxis that is deeply responsive and accountable to those it serves and a reimagined social science that calls for an engaged academic world that leaves the ivory tower and learns from experiences in community, organizational, and family settings. The next generation of research methods is emerging.

THE NEXT GENERATION: COMMUNITY-BASED ACTION RESEARCH

This chapter has focused largely on the philosophical foundations of community-based action research. I have presented arguments that seek to substantiate my assertion that it is a legitimate approach to research, despite procedures that vary from commonly accepted practices associated with scientific method. My intent has been not to show that scientific method is wrong or incorrect but to demonstrate that it sits alongside community-based action research as one of a variety of authentic approaches to inquiry.

Even within action research, there is a place for some of the methods, procedures, and concepts usually associated with traditional science. Insofar as people live in a physical universe, traditional scientific research methods—sometimes unhelpfully labeled "quantitative" research—can provide much useful information. We can describe, for instance, the number of people involved in a setting and how they are distributed geographically, organizationally, and culturally. We can also enumerate the number and proportion of unemployed people, the number and types of dwellings, the age distribution of the population, and relationships that exist among such features as gender, social class, race, educational attainment, employment, and poverty.

The meaning or significance of any of this information, however, can be determined only by the people who live the culture of the setting, who have the profound understanding that comes from extended immersion in the social and cultural life of that context. Numbers can never tell us what the information "means" or suggest actions to be taken.

I recently engaged in a conversation with a colleague regarding the utility of quantifying events. "Jim," I said, "if, over a period of six weeks, I read a local newspaper 40 times, have 120 meals, drink 30 bottles of beer, teach 24 classes, and punch the dean once, which event do you think will be most significant?" In this instance, it is easy to imagine which event is symbolically the most important. In many situations, however, outsiders, expert or otherwise, face grave risks in making judgments about the significance of events. Certainly, the number of times an event occurs is a poor indicator of its symbolic importance or the impact it is likely to have on people's lives.

Action research, therefore, ultimately focuses on events that are meaningful for stakeholders. It provides a process or a context through which people can collectively clarify their problems and formulate new ways of envisioning their situations. In doing so, each participant's taken-for-granted cultural viewpoint is challenged and modified so that new systems of meaning emerge that can be incorporated in the texts—rules, regulations, practices, procedures, and policies—that govern our professional and community experience. We come closer to the reality of other people's experience and, in the process, increase the potential for creating truly effective services and programs that will enhance the lives of the people we serve.

This move to emphasize the perspectives and agendas of the client group rather than those of professional practitioners in the development and improvement of programs and services does not mean that simplistic, populist ideas current in mass culture provide the basis for service delivery or institutional practice. For instance, the McDonald's fast-food curriculum model would provide a rather unhealthy educational diet in schools. It does mean, however, that professionals take seriously the ideas and perspectives of those they serve, developing ways of working that implicitly honor the intelligence and integrity of the people with whom they work.

Recent theoretical debates have provided substantial recognition of the philosophical strength of this approach to inquiry. The working principles and values revealed in the first chapters of this book are given credence by Denzin's (1997) forward-thinking work, aptly subtitled *Ethnographic Practice for the Twenty-first Century.* He seeks to engage social scientific practices that "will move closer to a sacred, critically informed discourse about the moral, human universe" (p. xviii). Drawing from the work of Christians, Ferre, and Fackler (1993) and Lincoln (1995), he advocates approaches to inquiry that engage a feminist, communitarian moral ethic organized by the following assumptions:

- Community as ontologically and morally superior to persons
- Dialogical communication as the basis for moral community
- Transformation as the major goal of any ethical or occupational practice
- Commitment to the common good and to universal human solidarity
- A sacred conception of science that honors the ecological as well as the human
- A stress on human dignity, care, justice, and interpersonal respect
- A belief that those studied are active participants in collaborative research processes, with claims over any research materials produced in the research process

This perspective is consonant with an ethic of caring (Collins, 1991) that celebrates personal expressiveness, emotionality, and empathy; values individual uniqueness; and cherishes each person's dignity, grace, and courage. It is a model, in other words, that fits the intent and processes of community-based action research. It emerges, philosophically, from *standpoint epistemologies,* that is, the ways of knowing and understanding implicit in the experience of particular social groups that challenge traditional approaches to social science. Standpoint epistemologies emerge from feminist theory contending that studies involving women should privilege women's lived experience and the standpoints, or perspectives, that experience brings to the ethnographic project. These perspectives challenge the notion of a single standpoint from which a final, overriding vision of the world can be written (Smith, 1989) and therefore question the ways traditional, patriarchal social science has been constructed. This dynamic extends to other interpretive communities and suggests that ethnic groups, people of color, gays, and so on draw on their group and individual experiences as the basis for texts that speak to the logic and cultures of these communities (Denzin, 1997).

Standpoint perspectives, therefore, advocate ways of formulating reports from the starting point of people's experience to provide a more accurate representation of reality, the production of local knowledge about the workings of the world (Denzin, 1989). Standpoint perspectives suggest the need to recover and bring value to knowledge suppressed by the existing texts and discourses of social science and by interpretations of people's lives that are inscribed in the apparatuses of the state, mass culture, and the popular media.

Patricia Hill Collins's (1991) Afrocentric feminist epistemology, for instance, focuses on the primacy of concrete, lived experience; the use of dialogue in assessing knowledge claims; the ethic of caring; and the ethic of personal accountability. "This epistemology creates the conditions for an existential politics of empowerment that will allow African American women to actively confront racial, gender and class oppression in their daily lives" (p. 237). Trinh (1992) writes from an Asian perspective, challenging systems of domination that seek to impose an essential identity on a person and her experience. She advocates the production of texts that allow women to experience and assert their difference in relation to others and to assume the position of active speakers, listeners, readers, writers, and viewers.

Although standpoint perspectives relate particularly to the experience of marginalized groups, there is a more general point. They should sensitize writers to the need to resist impulses to formulate objective, generalizable accounts claiming to include the perspectives of all people. Reports need to incorporate and represent the perspectives and experiences of the diverse social groups affected by the events and phenomena studied to ensure that the standpoint of each group is presented as authentic and legitimate, rather than an aberration from a hypothetical norm.

Not so long ago, the human services agency in my state produced a carefully compiled child protection manual, complete with detailed procedures for evaluating events related to the well-being of children and formulating appropriate intervention strategies. A consultant had carefully prepared the manual, taking into account the latest information on child protection available in the literature and using intervention strategies that were deemed to be appropriate. The procedures, however, were inappropriate to the realities of Aboriginal family and community life, applying criteria for child care that were contrary to the values and conventions of Aboriginal communities and advocating interventions that were reminiscent of more repressive eras in Aboriginal history. Aboriginal staff of the agency went on strike in one region, forcing the agency to have the offending manual reviewed and modified to take Aboriginal cultural values into account.

GIVING VOICE: REPRESENTING PEOPLE'S EXPERIENCE

The desire to give voice to people is derived not from an abstract ideological or theoretical imperative but from the pragmatic focus of action research. Its intent is to provide a place for the perspectives of people who have previously been marginalized from opportunities to develop and operate policies, programs, and services—perspectives often concealed by the products of a typical research process. Research reports often mirror the style of "objective" scientific reports, providing dry, abstract renderings of ideas and events and masking the fact that they are the productions of a socially situated author who writes from his or her history of experience. The process often encapsulates events within the mystifying jargon of academic disciplines or buries them in the conceptual apparatuses of a bureaucracy.

Previous sections of this book have suggested alternative ways for community-based action research participants to represent more effectively the outcomes of their research processes. The purpose of these alternatives is to ensure that accounts and reports capture the people's everyday, concrete, human experiences, allowing audiences to understand more clearly the realities of people's lives. The need for changes in reporting procedures is clearly presented in an increasing body of research literature (e.g., Berger Gluck & Patai, 1991; Bruner, 1990; Burnaford, Fischer, & Hobson, 1996; Lincoln & Tierney, 1996; Neumann & Peterson, 1997; Witherall & Noddings, 1991). These and many other authors explore ways of representing people's experience to more effectively grasp the many dimensions of life that formal, official reports often fail to reveal.

Denzin (1997) explores the underlying assumptions and principles of ethnographic writing and many of the dimensions of the reporting act. In so doing, he suggests that authors of official and academic reports might usefully employ technical processes used by journalists and fiction writers. He starts from the assumption that it is impossible to write objective accounts not ultimately inscribed with the perspectives and experiences of the author because of the interpretive processes inherent in data collection, analysis, and report writing. Drawing on a wide range of authors, he suggests that experiments in genre, voice, narrative, and interpretive style will provide more effective ways of knowing, other ways of feeling our way into the experiences of others. Such writing, he suggests, will reveal the meaning of events given by interacting individuals, focusing on experience that is deeply embedded in and derived from local cultural contexts that will include homes, offices, schools, streets, factories, clinics, hotels, and so on. These accounts and reports will recognize and represent the multiple interpretations of any event or phenomenon, yet also acknowledge the broader discourses that are inscribed or imprinted in people's local experiences—government policies, incipient racism, systems of privilege, imbalances of power based on social status, and so on.

Denzin works from the premise that the purpose of accounts derived from research processes is more than to record experience. Their goal, rather, is to provide the basis for transforming people's civic, public, and personal lives through the truths—the new ways of describing and interpreting events—revealed in the research process. Reports, therefore, must be more than impersonal, abstract accounts of "facts," which themselves are merely pieces of information acquired and interpreted selectively. They must be empathetic, evocative accounts that embody the significant experiences embedded in the taken-for-granted world of people's everyday lives. They must record the agonies, pains, tragedies, triumphs, actions, behaviors, and deeply felt emotions—love, pride, dignity, honor, hate, and envy—that constitute the real world of human experience.

In this context, the researcher-writer takes a different stance in the research and writing process. No longer "experts" capable of defining, describing, and interpreting the "facts" or "truth," researcher-writers position themselves toward writing processes that assist others in describing and interpreting their own experiences, sometimes acting as scribes for research participants or as coauthors or editors. The writer as scribe-for-the-other helps people give voice to their own interpretations of events, working with them to identify the key elements of their experience and shape them into a report.

CHANGING OUR WORK AND SOCIAL PRACTICES: "SCRIPTS" FOR POLICIES, PLANS, PROCEDURES, AND BEHAVIOR

Change is an intended outcome of action research: not the revolutionary changes envisioned by radical social theorists or political activists, but more subtle transformations brought about by the development of new programs or modifications to existing procedures. These developments and modifications, however, must necessarily be carefully planned and derived from the research processes to provide people with the means to more effectively deal with the problems investigated. The procedures presented in Chapters 6 and 7 suggest some ways of moving from reflection and theorizing to action. These moves need to be made more explicit so that the institutional imperatives of universities and bureaucracies do not inhibit the potential of the research process.

There is a need to more overtly review the outcomes of research so that they are presented in forms that may be applied directly to the texts that govern much institutional life: the policies, plans, and procedures that dictate how programs and services are delivered to the people served by those institutions.

Experiments with forms of writing that more effectively capture people's experience, discussed in the previous section, are moves in the right direction, but they appear limited in their ability to provide tangible outcomes for those who participate in research. Academic research tends to focus on the production of written texts that reflect the imperatives of university life, and research reports often remain trapped within the walls of the university. They circulate within the texts and journals of the university and, even when enacted as performances, are presented as theatrical productions playing to audiences oriented toward institutional life. They do not penetrate the everyday lifeworld of homes, schools, clinics, offices, shops, and factories and are especially remote from marginalized peoples whose lives are often framed by oral traditions that are misrepresented by the forms of understanding presented in written texts.

Even approaches to research designed to be more sensitive to these dynamics are prone to processes of knowledge production that colonize the experience of others. Denzin's (1997) interpretive ethnography, for instance, seeks to produce evocative understandings through a variety of literary genres that perform or write of people's experience (see Chapter 5). Research products, such as reports, presentations, and performances, are taken from the locations of the study and used as generalized narratives for an audience unrelated to the research context. Those texts and performances are used largely by others to enable them to understand the experience of those written about and often place that experience in an etic (outsider) framework of generalized theory. Such theorizing tends to be macrofocused, incorporating broad concepts, such as power, class, gender, race, and so on, often providing no basis for action in people's daily lives. The terminology turns the focus away from people's lived experience toward more general social processes that they have little opportunity or hope of controlling. Critiques within this type of literary production usually focus on negative constructions of the situation, often framed within descriptions and explanations of macrosocial and cultural structures and treated as real entities, rather than as constructed ways of interpreting social phenomena.

These texts and performances are sometimes illuminating, providing understanding of the broader contexts within which people's lives are played out, and revealing underlying forces of social life that are not evident existentially. They are, however, etic (outsider) constructs that have the power to distort people's lived reality and misrepresent the existential experience of their everyday lives. They tend to be generalizing interpretations that focus on underlying elements of social life, such as power, race, and gender, rather than revealing the dynamic, interactive play of people and events as they work their way through their day-to-day lives. Dealing with macrostructures as they do, they often fail to provide the possibility for restorative action and run the risk, therefore, of becoming voyeuristic and disempowering.

The use of those types of production should, perhaps, not be underestimated, because they do provide useful ways of communicating evocative interpretations of social life that inform readers of social realities far removed from their own experiences. Community-based action research, however, suggests the possibility of more socially responsive uses of research, providing the means for people to have a more direct impact on the significant issues that continue to detract from their social life and to make a tangible difference in the problems that diminish their lives. It reconceptualizes research as a participatory process that allows scholars—to follow West's (1989) dictum—to make their skills available for the purposes of the people.

Action research enacts localized, pragmatic approaches to research, investigating particular issues and problems in particular sites at particular moments in lives of interacting individuals and groups. Its purpose is to provide participants with new understandings of an issue they have defined as significant and the means for taking corrective action. The processes are necessarily participatory, enabling all people affected by the issue to have their voices heard and to be actively engaged in research activities. Action research suggests a move in emphasis from the creative texts of experimental interpretive ethnography toward the production of "practice scripts"—plans, procedures, and models derived from the final stages of action research that describe the actions people will take, or the behaviors in which they will engage.[1] Practice scripts—plans, procedures, and so on—also provide the means for people in professional or occupational roles to reformulate policies and practices to enable them to support the activities or changes in behavior of their clients, students, patients, or customers (see Box 9.1).

Although academic research often seeks critique from within the constructs of the social sciences, critique in action research derives from the conjunction or clash of the diverse perspectives of individuals and groups

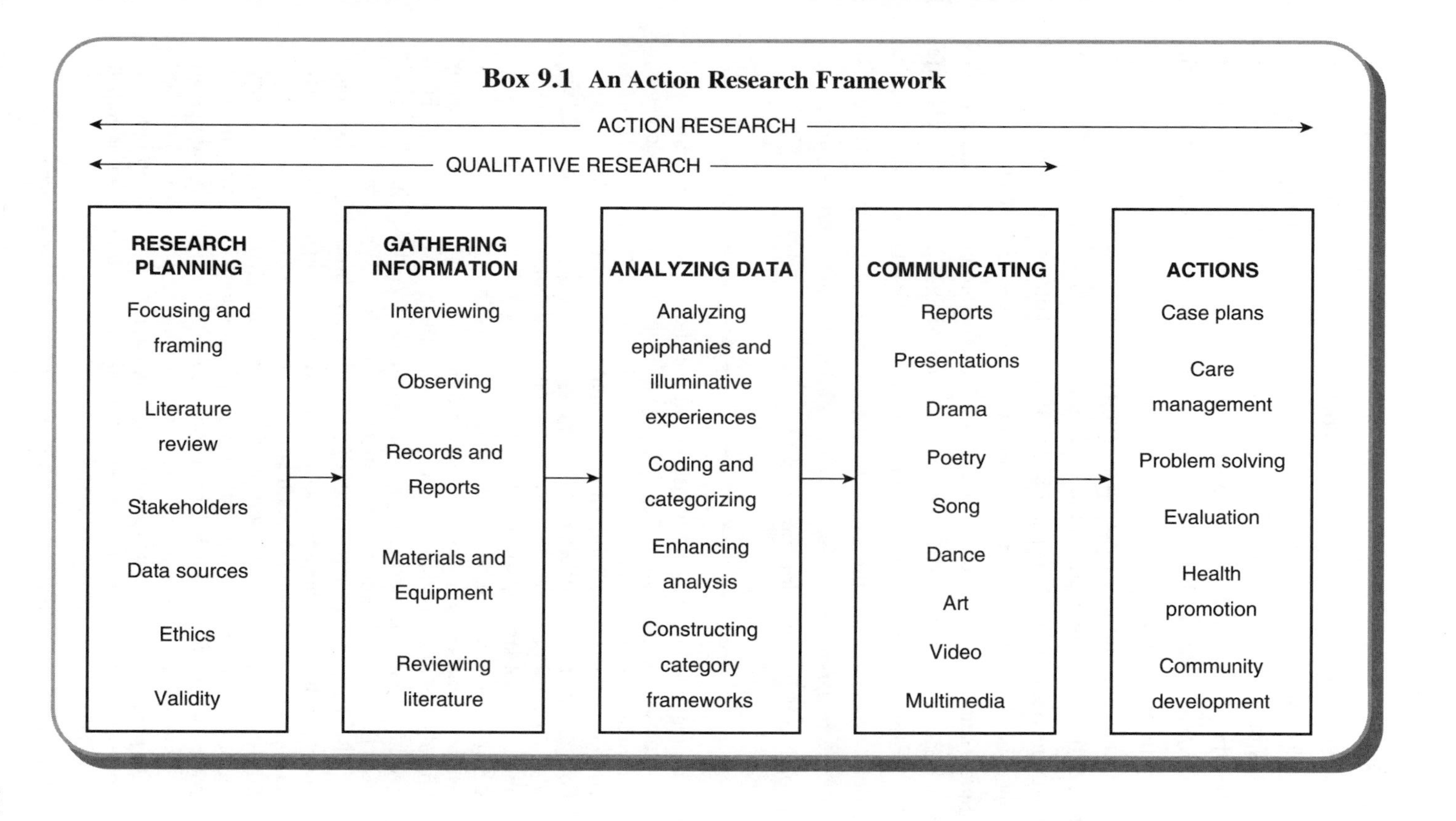
Box 9.1 An Action Research Framework
ACTION RESEARCH
QUALITATIVE RESEARCH
RESEARCH PLANNING
Focusing and framing
Literature review
Stakeholders
Data sources
Ethics
Validity
GATHERING INFORMATION
Interviewing
Observing
Records and Reports
Materials and Equipment
Reviewing literature
ANALYZING DATA
Analyzing epiphanies and illuminative experiences
Coding and categorizing
Enhancing analysis
Constructing category frameworks
COMMUNICATING
Reports
Presentations
Drama
Poetry
Song
Dance
Art
Video
Multimedia
ACTIONS
Case plans
Care management
Problem solving
Evaluation
Health promotion
Community development

participating, each perspective being challenged by the others. Critique is derived not from an etic, outsider interpretive framework but from the process of exploring and negotiating the different perspectives emerging from the process of investigation. It uses concepts and terms from within the context, rather than those drawn from academic texts or bureaucratic discourses, and seeks constructions and interpretations of the issue that provide the basis for direct, tangible uses within the research context.

Change, the desired outcome of community-based action research, therefore derives from the here-and-now ideas and concepts taken from the taken-for-granted lifeworld of the participants. It is based on the language, ideas, and concepts that make sense in their homes, offices, schools, factories, shops, clinics, centers, organizations, and institutions and are part of their everyday experience of family, community, work, professional, and commercial life. The products of research are not only written reports but also "practice scripts"—plans, procedures, models, maps, and so on—that provide the basis for reformulating practices, policies, programs, and services related to people's occupational or community life.

I am reminded of the long list of official policies, projects, programs, and institutional processes that have damaged the lives of minority peoples in Australia and the United States. I think of Dee Brown's *Bury My Heart at Wounded Knee* (1970) and Anna Haebich's *For Their Own Good* (1988), both of which depict the disastrous impacts of official policies on the lives of indigenous peoples. I think of the way that I organized my classroom in an Aboriginal school, so that children were forced to sit with others with whom they had a taboo or avoidance relationship. I think of academics who, from the safety of their ivory tower, formulate research projects that are implicitly critical of professional practitioners. The list goes on. Last night, a good friend and I, musing over these dynamics, thought of the old saying that "the road to hell is paved with good intentions."

But I am also reminded of the great triumphs of those who worked with people so that they became the instruments of their own liberation. Great leaders such as Mahatma Gandhi and Martin Luther King Jr. come to mind, but I am conscious of a flood of other local heroes with smaller accomplishments: the woman who established a play group to help young mothers come together to discuss their common issues—the genesis of a widespread network of play groups in my own state—and the psychiatrist who worked with parents and students in her local community to establish an independent school as a haven from a school ruled by an authoritarian principal. Fortunately, *that* list also goes on.

IN THE COMPANY OF FRIENDS

In this chapter, I have spoken of the dimension of power in community-based research. For some, the move to empower people is seen as a set of divisive, confrontational interactions and activities that enable one person or group to wrest control from another. The communitarian ethic embedded in action research, however, suggests a more consensual approach to investigation requiring participants to enact the working principles discussed in Chapter 2. The position of the researcher changes from one of controller, advocate, or activist to one of facilitator. The researcher's role is not to push particular agendas but to neutralize power differentials in the setting so that the interests of the powerful do not take precedence over those of other participants. Interactions and activities must, however, allay the fears of people in positions of power, generating levels of trust that enable them to feel sufficiently at ease to release some of their control over events. Their acquiescence is an essential feature of community-based action research. They should feel that their institutional or organizational responsibilities, or their place and status in the context, are not threatened by the research activities.

Procedures must also generate trust within less powerful groups, so that they are willing to participate in arenas in which previously they have been mistreated or demeaned. The reward for both types of groups is that they acquire ends or goals that otherwise would not have been possible for either, working by themselves.

The power broker role of the researcher is exercised with the intent of facilitating change in the power dynamics of the situation. The result is not so much a transfer of power from one group to another—although that often occurs—as a change in the *nature* of power relations. Power is still exercised, but in a different way. A teacher in a classroom does not relinquish the responsibility of facilitating students' education. The teacher can, however, choose to provide significant decision-making opportunities to students regarding the organization of the class, the curriculum, the time table, and so on. Social workers and health professionals may similarly engage client groups in the process of formulating or restructuring the ways in which programs and services are organized and administered.

Improved service delivery and problem resolution are not the only goals of these types of activities. Action research seeks to formulate ways of living and working together that will enhance the life experiences of the participants.

It provides ways of working that counter the impact of the current overemphasis on the technical and the bureaucratic and enables the formulation of procedures that take into account those features of a situation that are essentially human—that speak to issues of emotion, value, and identity.

Ultimately, however, the routines of community-based action research suggest ways of working that enable a harmonious and productive sense of social life. The end product is, if anything, peace of mind or the chance to participate in the pursuit of happiness. It is a movement away from competitive, power-driven, conflict-ridden organizational processes toward more cooperative, consensual ways of living.

A colleague once suggested that education could be defined as "the search for truth in the company of friends." Although "truth" may elude us, community-based action research might, in a similar vein, be defined as "the search for understanding in the company of friends."

Reflection and Practice

1. Reflect on the issues that have been presented in this chapter.
2. Identify concepts or issues that you find most interesting or informative.
3. State how or why you find them interesting or informative. How are they meaningful for you?
4. Are there sections of the chapter that are uninformative or unclear, or that do not seem relevant to your understanding of action research?
5. Are there are other theoretical positions you know that might usefully have been included that would extend people's understanding of action research?
6. Summarize the main points made in this chapter. State the strengths and weaknesses of the chapter, from your perspective.

NOTE

1. The notion of a "script" is drawn from those used for plays or films. Not only does a script carefully define what takes place, but it also delineates what will be said, and how things will be done and said.

BOX 9.2

Theory in Action Research

Purpose

To understand the way theory relates to the practice of action research

Content

The Place of Theory in Action Research

Theory in action research has a different position and purpose than theory in quantitative studies.

Theory and Practice

The outcomes of action research are increased clarity and understanding that provide the basis for resolving the problem on which the study focused.

The Legitimacy of Action Research

The legitimacy of action research comes not from the criteria associated with quantitative research but from its ability to be meaningfully applied to problems and issues in people's everyday lives.

Power and Control

Postmodern perspectives show how attempts to control people's lives are confounded by the creative, dynamic construction of social life and behavior. This points to the need for democratic, creative, and liberating ways of conceiving and organizing human activity.

The Next Generation

Action research suggests the need to liberate and empower people through collaborative and caring processes of investigation and action.

Representing People's Experience

Action research seeks to produce empathetic, evocative accounts that embody the significant experiences embedded in people's everyday lives.

Scripts for Professional and Social Practices

The outcomes of action research provide "scripts" to guide new ways for enacting events and activities.

In the Company of Friends

The end product of action research is a harmonious and productive sense of social life. It seeks to move people to cooperative, consensual ways of living that provide peace of mind and contribute to the pursuit of happiness.

APPENDIX A

Case Examples of Formal Reports

The two reports in this appendix are different in style but follow the basic procedures for formulating a report outlined in Chapter 8.

Note that the first report, "Transitions," contains a report within a report. Part of the formal report is a report presented by the author to the women with whom she had been working.

Transitions—The Experiences of Older Women From Hospital to Home

Deborah Rae

INTRODUCTION

This report outlines an action research process that was conducted with women from the Older Women's Network in Mackay. Action research is a methodology that recognizes that the researcher, as a fellow human being interacting with others within a social context, is necessarily an implicit part of the research. It is also a cyclical, reflexive process that advocates continued learning and development. I have prepared this report in the first person, which illustrates the transparency of my subjectivity in this process and the benefits of this subjectivity and also allows for my ongoing reflection on my actions throughout the action research process.

1. PURPOSE OF THE REPORT AND THE INTENDED AUDIENCE(S)

This Report Has a Dual Purpose

One of the purposes of the report is as an assessment item for SWSP7348 Community Based Action Research as a course in my study of a master of social administration at University of Queensland. It is intended to demonstrate my use of a participatory action research process in an activity within the particular community in the Mackay (north Queensland) district. As such, its intended audience is the lecturer for this course, Dr. Ernie Stringer.

Within this report is another report, which details the progress of this action research to date. As such, its intended audience is the primary stakeholders and their identified community for this research. At this stage, this includes older women who have had experiences in transitioning from a hospital stay to home (primary stakeholders) and the Older Women's Network, of which these women are all members (identified community).

2. ISSUE INVESTIGATED

After a preliminary inquiry of possible issues or problems was conducted in late August 2005, the following issue was raised, with the understanding that it could possibly change after further investigation and clarification with a broader range of stakeholders.

> Issue: Older women are experiencing problems when they return home following a stay in hospital, such as not having the capacity to cook meals, perform routine household tasks, or care for other family members (e.g., an older spouse).
>
> Problem: There are no transition services from hospital to home for older women.
>
> Question: How do older women experience the transition period from hospital to home?
>
> Objective: To understand what this experience of the transition from hospital to home means for older women.

While this report is based on the previously described issue and objective, following the focus group meeting on October 20, modifications were made to these. These modifications are detailed in the following section titled "Outcomes of the Study."

3. CONTEXT OF THE STUDY

Mackay is located on the Queensland coast, 1,000 km north of Brisbane, and covers an area of more than 90,000 km^2, equivalent to about 5.2% of the total area of Queensland (Mackay City Council, 2005). In 2001 it had an estimated population of more than 143,578 permanent residents, of which approximately 69,800 are females. More than half of the total population (78,401) live within the jurisdiction of the Mackay City Council, while others live in outlying districts. Approximately 11,935 people in Mackay are over the age of 65, which means that Mackay's population is younger than that of the rest of Queensland.

However, this does represent an increase of 12% in the Mackay elder population from 1996 to 2001 (MWREDC, 2003).

The region is serviced by the Mackay Base Hospital and Mater Misericordiae Hospital and to a lesser extent by the smaller Pioneer Valley Private Hospital. The Mackay Base Hospital, which is publicly funded, provides a range of medical facilities and services, with other services provided by visiting specialists from Townsville or Brisbane. This hospital has recently been the subject of media reports concerning a lack of human resources, especially a lack of appropriately trained medical specialists, a lack of community confidence in its services, declining workplace morale, and workplace bullying.

Other key community health services in the Mackay region include Blue Care (in-home nursing care), Meals on Wheels, Golden Doves (care for people with disabilities), Home Assist (general handyman services), and Home and Community Health Assistance, including a local bus service for older and/or disabled people.

Researcher/Historical Perspective

Because I have no previous professional experience in community development and have been a member of the Mackay community for only a short time, the context of my research has included

(a) Making initial contact with an appropriate organization and key people within the organization
(b) Negotiating entry into the organization
(c) Identifying further opportunities for other individuals to enter the research group to participate in a possible research project
(d) Making contact and negotiating with key stakeholders external to this organization to initiate a research project

Stakeholders, Sites of Action, and Time Frames

Primary Stakeholders

Originally, older women who have had experiences in the transition from a hospital stay to home were the primary stakeholders in this action research project. These stakeholders were initially drawn from the Older Women's Network (OWN), which is "an independent forum in which the special needs of older women could be specifically addressed" (OWN, 2004).

At the focus group meeting on October 20, these primary stakeholders decided to modify the context of this action research to include both women and men. They also decided that being identified as "older" is not a numerical representation, but subject to self-identification by the person. For example, more explicit self-identification as an older person is represented by a decision to participate in social groups such as the OWN, 50 and Better, Independent Retirees, the Senior Citizens Association, or National Seniors. They also decided that the context would include older people who have had an experience in any of the private or public hospitals in the region and that the project would not be limited to Mackay city limits but also its immediate regional districts. This involves western, northern, and southern districts within approximately 50 km from Mackay city limits.

Other Stakeholders

- During interviews with individual primary stakeholders, when the project was focused on the experiences of older women, other possible stakeholders for this action research were identified. During the October 20 focus group meeting, however, these stakeholders were further clarified by the primary stakeholders and divided into separate groups. They now included:

(a) Other primary stakeholders
  - Men and women from organizations such as Independent Retirees, Older Women's Network, 50 and Better, the Senior Citizens Association, National Seniors, and Australian Pensioners League
  - Friends and acquaintances of current primary stakeholders

(b) Hospital representatives
  - Representatives of Continuum of Hospital Extended Care (CHEC), which "provides care and support for discharge from hospital"
  - Social workers

(c) Community service providers
  - Home and Community Care (HACC) transport services
  - Blue Nurses in Mackay
  - Meals on Wheels in Mackay
  - Home Assist in Mackay
  - Dial an Angel (based in Southeast Queensland)
  - RSL Home Care in Mackay

(d) Other community members

- President of the Mackay Regional Council of Social Development (MRCSD), who is also a local member of Queensland Council of Social Services (QCOSS)
- Member of the Mackay and District Health Council
- President of National Seniors, who is also on the State Council for Seniors

Sites of Action

To date, the project has essentially been situated in locations that were preferred by the primary stakeholders. Specifically,

- Meeting with OWN at their usual venue (Blue Nurses function room)
- Meeting with primary stakeholders in their homes (except for one who chose to meet at a coffee shop)
- Focus group meeting at a primary stakeholder's home, as decided by the stakeholders themselves

Because the primary stakeholders are expecting the size of the meeting group to expand, they have commenced inquiries into accessing an alternative meeting venue. Options they are currently considering include the MRCSD's meeting room, the Mackay Neighbourhood Centre, or the Mackay City Council library.

4. PROCEDURES USED TO CARRY OUT THE STUDY

(a) Data gathering

Data gathering has been accomplished by

- Participation in meetings of the OWN, to gain access to potential primary stakeholders
- Identifying people I could approach to participate in the action research project, as this group is comprised of many primary stakeholders. This included identifying opportunities for identifying people with particularly tragic experiences (extreme case sampling), usually reported experiences (typical sampling), particular knowledge of this

issue (concept sampling), and divergent experiences (maximal variation sampling; Stringer, 2004).

Comments by individuals at this meeting seemed to suggest that they are particularly interested in achieving a specific tangible outcome that they have already thought about. I attempted to be mindful of this when I met individuals and heard their stories, so that

- They had an opportunity to fully explore their own experiences and possibly come to realize a range of issues and possible options
- They had an opportunity to benefit from the process, not just the outcome

This required me to use grand and mini-tour questions effectively, make full use of stakeholders' opportunity to review and clarify their words, and give stakeholders plenty of time to reflect on their actions and statements.

Data-gathering activities will ensure that primary participants remain included in, and the focus of, the process, as well as the determining elements of the "substance of our research." Participants' involvement in contacting other stakeholders and arranging and chairing future focus group meetings is an initial step toward guarding against the dangers of excessive networking, neglect of primary stakeholders, and the predominance of the "agendas and perspectives of organizational people."

Clarifying My Role

Even though I explained that it is not my role or intention to lead the project, or do all the work, it appeared that OWN members were still assuming that this was indeed the case. For example, an individual asked if I expected the outcome would be something that would be government or privately funded. I therefore think that, although I may have initially explained my role in the process, it will require further clarification, probably on a few occasions. Also, I think my actions may not have been consistent with my words. For example, I was introduced by the president while standing in front of a seated group. I believe this created an impression that I had some sort of authority over the process. It would therefore have been better for me to initially meet the participants individually, rather than in this group forum. This could have been achieved during informal chats during the morning tea break (which was my original intention).

Note Taking

Note taking initially proved to be difficult because of my focus on verbatim recording of each participant's words. This detracted from my ability to keep eye contact and maintain some degree of natural conversation with the participant. This proved to be less of a problem in later interviews, however, when I changed my physical position so it was easier to maintain eye contact, and my note-taking skills began to improve, so that I could relax a little and give more attention to the participant. I also found that in the pauses while I was writing, the participant had an opportunity to contemplate what she had been saying, which was reflected in her subsequent comments.

Participant Reactions

Judging by the passion and intensity with which these stories were told, it would appear that participants generally appeared to appreciate the opportunity to be heard and to tell their story. I attempted to ascertain their perceptions of the process by inviting comments about the interview or directly asking them for their comments.

Focus Group

I conducted a focus group with the primary stakeholders. The preparation, structure, and format were as follows:

> The meeting time and date were scheduled by two of the participants, who then telephoned me and some of the other participants, while I contacted two others.
>
> The meeting was held in the home of one of the research participants.
>
> The meeting was attended by six primary stakeholders, three of whom have been interviewed once; the other three have been interviewed twice.
>
> All participants were provided with transcribed copies of their interview(s).

The meeting duration was 2.5 hours. Participants requested that I chair the meeting, "since it's the first one."

Focus Group Outcomes

The meeting was, from my perspective, very successful. This was evident from the outset, when the passion and depth of these women's concerns became evident before the meeting started. This suggests that the timing of the group meeting was appropriate, as they are all currently keen to share their stories and make this issue public. The tone of the meeting was informal and friendly, but also focused on a particular objective. It was lively, enjoyable, and very active, but also primarily concerned with the issue at hand. The meeting provided an ideal forum for these women to express their passion about this issue and also actively work toward realistic options for resolution. Similarities can be seen with Stringer's description of a workshop, where "the level of animation in their discussions and the extensive lists of useful information emerging from their discussions were testament to their enthusiasm and the extent to which they appreciated opportunities to learn from each other" (2004, p. 123).

Artifacts

A range of brochures and pamphlets were reviewed by other participants and me throughout the interview and focus group process. These were predominantly documents related to hospital preadmission and discharge processes and services, medication sheets, and newspaper clippings about community services related to transition from hospital to home. The data within these documents were useful in developing the knowledge of participants regarding this issue, particularly in enlightening us about the range of services available. A critique of this data also revealed how its stated intentions did not match reality, which was emphasized by participants in their interviews.

(b) Data Analysis

Data analysis in this action research project has been conducted according to the interpretive interactionist methodology of identifying and analyzing epiphanies and illuminative experiences. As Denzin (2001, p. 158) defines it, an epiphany is a "moment of problematic experience that illuminates personal character and often signifies a turning point in a person's life." It is within the research participants' epiphanies that the researcher may "capture the concepts, meanings, emotions and agendas that can be applied to problems affecting their personal, institutional, and professional lives" (Stringer, 2004, p. 99). Data analysis processes therefore include

Identification of epiphanies in data collected in the first interview with individuals

Deconstruction of these epiphanies to reveal their features and elements

Preparation of accounts and narratives of these epiphanies for the individuals

Preparation of a joint account, via the focus group meeting of primary stakeholders

Interpreting the participant's world

I had significant concerns about using the data gathered from participants in any way, as I feared that I should distort their perspectives. I considered it more important to keep their words and concepts intact, without any interference on my part. However, after reading Denzin's (2001, p. 133) explanations of the "thick, contextual, interactional, multivoiced interpretation," it became evident that this actually provides the reader with an opportunity to extend his or her understanding of the issue being investigated.

(c) Report Construction

Following the focus group meeting, it was agreed that I would prepare a written account from the meeting minutes of the project to date, including the issues raised by participants. Participants also agreed that they would prepare a verbal account to present to the OWN members. My written account, which served as a progress report for OWN members, is provided in the later section titled "Outcomes of the Study."

(d) Ethical Procedures

At the commencement of the project, ethical issues to be addressed included

Maintaining ethical principals of MRCSD

Maintaining the confidentiality of interactions with participants, particularly with respect to any comments they provide about local services

Ensuring appropriate storage of confidential information at MRCSD

Obtaining permission from participants to share with others information they provide, and how it will be shared (e.g., verbally in general references, in written brochures, etc.)

(e) Checks for Rigor

Checks for rigor within this action research project include the following:

- Prolonged engagement

Individual interviews with participants (either one or two) were each at least two hours in length, and this prolonged engagement enabled sufficient exploration of the issue at hand. This was coupled with the initial interaction with primary participants at the OWN meeting.

- Persistent observation

The initial meeting, first and second rounds of interviews, and the focus group meeting took place over a period of approximately 10 weeks. This element also included sending e-mails and making telephone calls throughout this period.

- Triangulation

I used a range of sources, including individual interviews, focus groups, participation in a group meeting, observation of interactions between participants, literature review, and a review of artifacts, including brochures and other documents.

- Transferability, dependability, and confirmability

The thickness of the detail of each individual interview is evident in the transcribed documents, and its applicability to this whole group was confirmed in the focus group meeting. That is, issues of each person are clearly identifiable in her accounts, due to the level of detail, passion, and intensity in participants' stories. To date, the primary participants seem assured of the dependability of this data, which has been fully transparent and provided to them at every opportunity. This dependability will be assessed more rigorously with the production and distribution of the progress report to the OWN meeting. The confirmability of the data is clearly apparent even within this report, as all data, including electronic and paper meeting minutes, learnings reports, brochures, and other documents have been filed and stored.

- Philosophical rationale

Action research is a form of naturalistic inquiry with the purpose of studying people's subjective experience and "explor[ing] perspectives on an issue or problem" (Stringer, 2004, p. 16). Thus it is a form of "transformational learning" (p. 3), which provides research participants with the opportunity to gain enhanced clarity and realizations of their own situations within a social context. Action research processes therefore include a progressive development of events, through continuous cycles of looking, thinking, and acting. Simply, it is research that is conducted by, with, and for people (Reason & Bradbury, 2001, p. 2). Reason and Bradbury (2001, p. 1) also emphasize the democratic nature of action research that is concerned with "developing practical knowing in the pursuit of worthwhile human purposes" with a view to "the flourishing of individual persons and their communities." This occurs within the realm of the interconnected complexity of people and their cultural worlds and the expectation of outcomes that are perceived as worthwhile by these people in their everyday lives. Action research is therefore concerned with "a more equitable and sustainable relationship with the wider ecology of the planet of which we are an intrinsic part" (Reason & Bradbury, 2001, p. 2).

OUTCOMES OF THE STUDY

The outcomes of this study are represented in the following progress report that will be forwarded to the OWN membership. While this report will provide them with a written record, the research participants will also make a verbal presentation to the OWN meeting of research outcomes to date. This report has been prepared by me at the request of the research participants and is, in essence, the minutes of the focus group meeting of the primary stakeholders. Its framework has been constructed according to the key elements identified by the research participants' joint account of this issue.

Transitions: Experiences of Older People From Hospital to Home—Progress Report

INTRODUCTION

This is a report to the members of the Older Women's Network about the progress of an action research project being conducted by some OWN members with Deborah Rae, a volunteer worker with the Mackay Regional Council for Social and Regional Development (MRCSD).

This progress report includes

- A brief description of the background of the action research project
- The project's objective
- A summary of the key issues that OWN members identified
- A conclusion
- Background

The OWN members who have been a part of this research have

- Met with Deborah for an individual interview, where they talked about their experiences when they went from hospital to home
- Met with Deborah again (some, but not all participants) so she could verify what she told them and so that they could explain their story further
- Met together as a focus group to share their stories and decide what to do next

OBJECTIVE

The issue that OWN members have been talking to Deborah about is that older people, especially those who live alone, experience problems when they return home after a stay in hospital, such as not being able to cook meals, do housework, or care for other family members (e.g., an older spouse). The problem is that there are no transition services for older people from hospital to home.

The OWN members' objective is to find ways to improve the experience of the transition from hospital to home for older people.

SUMMARY OF THE KEY ISSUES

It Could Be Anyone—It's a Community Issue

The participants feel that they shouldn't just do this for women. It should be "across the board, not just channelling down one lane to one sort of people." It's a community issue and "should be available for anyone that requires it." One woman explained, "The community as a whole needs to say, we gotta put something in place. Everyone needs to know they've got someone to fall back on, if they need it. It's about that support. It's not a luxury. It's a basic need for the community. It should be there for people of any status."

Infections

Participants talked about how you can become very sick from infections after you get out of hospital. One woman explained, "If you're lucky, you don't get an infection. The infection actually caused more hassle. The infection part is one of the really major problems that the hospital has got. And the patient has got. The hospital just can't fight these germs anymore."

Another very independent woman had a major operation that was very successful. She didn't know that she had an infection until her daughter found her collapsed in her bedroom. She said that infections "take over you—you have no control. It just knocks you for six. If you don't have a positive attitude, the infection can just take control. You're not aware of it yourself. You don't realise you're getting sicker. Then you're unable to get up. You just collapse and stay there. Somebody on their own wouldn't have someone to push them."

Who's Responsible for Telling You This Information?

Many women talked about problems with the information they get from the hospital and from other places. They also talked about how "you've got to make people aware, educate them." They said that you "find things out when you don't need it anymore," or "they give you information but then you can't access it," and "you hear people talking about these things [services]. That doesn't mean that when you come to need it, you get it right." Even during the focus group meeting, some women found out about some services for the first time, exclaiming, "I didn't know you could do that!" "Yeah, I just found out too."

One woman talked about how she saw her friends struggling to cope with finding information. She said, "I've seen women who are so depressed. You say hello and they fall to pieces. They're lost as to who can help them and who can't. They suffer with depression. They don't know what to do in this situation. They can't help themselves. They don't know how to get *into* these services. I know that I've had to resource things myself to find out what is available."

One woman was given information in a brochure about a service that would come to her house to help her. But later she was told that they wouldn't do it. She says, "It's about having things at your fingertips when you need it. They should be saying, 'This is what we can offer you now.' It could have happened at preadmission. It's all about the timing. You need appropriate information that can be followed through. My life could've been made so much easier."

How Am I Going to Get There?

Some women talked about the problems with public transport in Mackay, which is a "problem when you come out of hospital, and have to go back a few days later." Many women don't drive: "Public transport is almost non-existent, depending on the area," and "taxi transit is such a hassle. If you're sick you can't be bothered with all that."

One woman said that after a successful operation on her neck, her main distress was when she was told to go to her doctor to get the staples out. She explains, "I was feeling exceedingly unwell. I had to get the staples out. I had nine of them in my neck. It's hard to get to a doctor. I rang all day. I was exhausted. I just kept thinking 'Where in hell am I going to get these staples out of my neck? How can I *get* there?'"

You Need Someone to Call on You

Several women said that they were most concerned about women who are living on their own. One woman, who lives alone herself, explained that "if you're living on your own, you need to know how to arrange tradesmen, pay rates. But if you lose your husband in your sixties, you're thrown in the deep end. You have to learn all these things, especially if they have no children who have time to help them out."

Another woman, who lives in a group of units, has seen many friends and neighbors struggling to manage alone. She says, "The Blue Nurses get sent out, just to redress the wound. They're not there all the time. Even Meals on Wheels don't have time to stop and chat. They rush in, drop the meal and go. They have to struggle to get around at home, even to make a cuppa. For dinner they just have a packet of biscuits and a cup of tea. It's not enough. Some people wouldn't even bother with that. They just sit."

I'm Happy to Be Home, But Is It Practical?

One woman, who has lived at home alone for many years and enjoys her independence, explains, "There are a lot of hazards when living alone at home, and you do things you shouldn't do. Old people and children don't know their limitations. I think I can do that—then find you can't." The group talked about the dangers of having an accident, and how it changes your life. As one woman who lives alone explains, "At our age, if you get a fractured hip, half your life is over. It takes a long while to heal, if it heals, and your whole life is altered. Its structure is altered. I'm scared as hell of slipping over."

They also talked about the ability to plan ahead if you live at home. For example, as one woman says, "If you have an accident at home, then you're in an ambulance, in old clothes, they've ripped the leg of your pants to look at your knee . . . You haven't had a chance to put anything in an overnight bag . . . How do you handle that? You have to ring a friend and explain where everything is in your house. In the end of my wardrobe, I have an overnight bag with nightie, slippers . . . You have to be prepared. You have to realise these things."

FAMILY AND FRIENDS

Many women talked about having family and friends who would "put themselves out for me," but were not keen on "family having to interrupt their lives." One woman who developed an infection and suddenly became very ill explains, "When I came home I was sicker than when I was in hospital. My daughter had to leave her two children and come down from Townsville to look after me. It was very, very inconvenient. I was lucky that I had that option. Not everybody has that."

They also talked about feeling pressured to get support from friends. One exasperated woman exclaimed, "Friends, friends, friends, friends! Well if I kept asking friends, I won't have any left. I didn't feel comfortable asking for help. Yeah, I felt *really* uncomfortable with it. I had to do what was asked of me [in preadmission]. I was totally pressured. I absolutely had no choice—you were pushed onto your friends. I think there should be an alternative."

All the women said they would help their friends and family, but some admitted that they preferred to manage alone when needing help themselves. One woman reveals, "I should've [called someone to help] but . . . A neighbour came over to see if there's anything I needed, but I said, 'Oh I'm alright.' Getting around I could manage it, it just took a long time. I didn't want anyone to take me to the loo. I'm one of those silly ones—too independent. I've been living on my own a long time."

CONCLUSION

The OWN members participating in this research have identified several key issues for older women in their transition from hospital to home. They have also started to find some ways to improve this transition and have an action plan in place to continue researching this issue and working toward possible improvements. They have decided to meet again in November and will keep OWN informed of their future actions.

LITERATURE REVIEW

To date, the research participants have not expressed an interest in reviewing literature related to this project. They are aware that similar research has been conducted by OWN members in other states (although not based on action research methodology), but at this stage they have made the decision to continue gathering data from older women within their own organization, as well as to broaden their data gathering to women and men from similar organizations in Mackay. Simultaneously, they have requested that I commence gathering data from other stakeholders, including community service providers and hospital personnel.

The data in this review is derived from three sources:

A health and ageing submission to the federal budget, prepared by the Council of the Ageing (Australia)

A survey on experiences of care after hospital, conducted by the Older Women's Network (Action), Inc., group in Canberra

A research activity describing the "reported experiences of elderly patients regarding their transition from an acute hospital to independent community living," as reported in the *International Journal for Quality in Health Care*.

REFERENCES

Denzin, N. (2001). *Interpretive interactionism* (2nd ed.). Thousand Oaks, CA: Sage.

Mackay Whitsunday Regional Economic Development Council (MWREDC). (2003). *Mackay Whitsunday regional statistical profile.* Retrieved from www.mwredc.org.au

Older Women's Network, Mackay. (2004). *About the Older Women's Network.* Unpublished brochure.

Reason, P., & Bradbury, H. (2001). *Handbook of action research.* Thousand Oaks, CA: Sage.

Stringer, E. (2004). *Action research in education.* Upper Saddle River, NJ: Pearson Education.

A New Mathematics Curriculum

Karen Swift

THE ISSUE TO BE RESEARCHED

At Blacksmith Primary School I am a member of the mathematics curriculum team and have a real interest in this subject area. Over the past two years, this team has been looking at the Curriculum Frameworks in the mathematics area. We have been investigating and identifying the key points and strategies of the learning area outcomes. We have inserviced our teachers and informed them of the information we have learnt from our investigations. We have also discussed with the teachers new strategies and techniques they should be using in their classrooms to achieve the outcomes stated in the Curriculum Framework document.

As it has been more than a year since the changes have been implemented, I was keen to see how the teachers, parents, and students of the school were experiencing the new changes. I had intended to complete this research project on the whole school but due to my inexperience as a researcher, the short allocation of time to conduct the research, and my determination to start small, I decided to complete my research project just on the five junior primary classes (Year 1 to Year 3) within the school.

STATING THE RESEARCH QUESTION

How are the teachers, parents, and students in the junior primary classes (Year 1 to Year 3) experiencing the new mathematics curriculum?

METHOD

After deciding on the research problem I discussed with the school principal my research proposal. She thought it sounded very interesting and was keen to see the results in the final report and presentation at the completion of the project.

I began this research project by purposively selecting a sample of participants from each of the groups significantly affected by the issue being studied—achieving mathematics outcome—taking into account gender, age of participants, academic ability, and ethnic backgrounds. The participants who were included consisted of three teachers, three students, and three parents from all of the junior primary grades. Once the students had been selected, I wrote letters to the children's parents explaining the project and asking for their permission for their child to be included in the project. I also gained informed consent from the other participants involved.

All the participants were interviewed individually. The teachers and parents were interviewed over a two-week period in my classroom before school or straight after school. The children were interviewed in their own classroom so they could give a grand tour and explain their own classroom procedures. I used ethnographic interview questions (Spradley, 1979), and I recorded the participants' responses using field notes. I then spent a week analyzing the data from each individual interview. I decided to use the categorizing and coding method of data analysis. I photocopied my field notes and then cut up all the words into little units of meaning. I then placed the units of meaning into categories, thus enabling the participants to interpret and have a better understanding of the data collected.

Once I completed all the initial interviews with participants, I decided to have a focus group meeting with the three different groups. These were short meetings during the last week of the school term. The aim of these meetings was for all the members involved to get to know each other, to be informed about the next procedure, and to decide on a convenient time to meet each week to work collaboratively to investigate how we could achieve mathematics outcomes in the school. The teachers decided to meet on Tuesday afternoons, the children at Wednesday lunch times, and the parents on Thursday afternoons.

When I completed my data analysis from the individual interviews I returned the analyzed data to the participants for member checking. For the students, I individually read their report with them.

The participants then gathered together for another focus group meeting. During these focus group meetings the participants created joint written accounts that gave a clear understanding of how all the participants in their group were experiencing and perceiving the issue.

After all the focus groups had created their group reports, a meeting was held where all the participants attended. At this meeting each focus group presented its report. The teachers used a PowerPoint presentation, the parents wrote a letter addressed to the teachers, and the students created a poster that expressed their likes and dislikes for math. The teachers also presented the information they had discovered during their literature study. After the students presented their report they left, and the teachers and parents worked collaboratively to create a narrative report to present to the principal and teaching staff. To create this report, they broke up into pairs of a teacher and a parent working together to identify common themes from the three reports. Each of the pairs reported back their findings, and then a group report was constructed.

During the next meeting the parent and teacher participants created a plan of action. The group identified the key issues to be addressed and suggested solutions to accomplish the desired outcome. An action agenda plan was then established using a planning chart. Next, the action agenda plan was added to the narrative report that was presented to the principal and teachers at a staff meeting.

The teacher and parent focus groups continued to gather once a week during the fourth term to continue to supervise and monitor the progress of the actions being implemented during that term.

MY PHILOSOPHICAL RATIONALE

An experimental style of research is not appropriate for my research issue, as I am looking at the behaviors and attitudes of teachers, students, and parents in relation to the mathematics curriculum. According to Stringer, "The social and behavioral worlds cannot be operationalized in scientific terms because the phenomena to be tested lack the stability required by traditional scientific method" (Stringer, 1999, p. 192).

The naturalistic inquiry approach studies people's subjective experiences and explores their perspective and views of a particular issue or problem. By using this approach, researchers "gain greater insight into the ways people interpret events from their own perspective, providing culturally and contextually appropriate information assisting them to more effectively manage problems they confront" (Stringer, 2004, p. 15). Once the researcher gathers the information from the people's experiences, he or she uses interpretive methods to analyze the data before producing reports that give detailed descriptions of the people's understandings of events and their behaviors. The researchers' reports and presentations can vary depending on the purpose and audience involved in the research.

Action research is at the center of the naturalistic qualitative research approach, as there is a "need to understand how people experience and make meaning of events and phenomena" (Stringer, 2004, p. 26). Action research "is a cyclical, dynamic and collaborative process in which people address social issues affecting their lives" (p. 4).

Action research uses a number of cycles of the Look-Think-Act sequence. During the Look or Observation stage the researcher gathers the information from a variety of sources and builds a picture to clarify the nature of the research problem or issue. The main information is gathered through interviews with stakeholders, where each stakeholder gives meaningful accounts of their personal experiences and perspectives of the research topic. Researchers also use participant observation, surveys, artifacts, and literature reviews to gain a clearer understanding of the problem or issue. Artifacts can include documents, records, student work samples, materials, equipment, and facilities.

The second part of the cycle, the Think or Reflection stage, is where the researcher analyzes large amounts of data to reduce the quantity of information and to gain a clearer understanding of what is happening and how it is happening for each stakeholder in relation to the problem or issue being investigated. According to Stringer (2004), there are two ways of analyzing the data, which are (a) analyzing epiphanies and (b) categorizing and coding.

Analyzing epiphanies involves the researcher identifying the epiphanies from each individual interview with the participants. An epiphany is an "illuminative or significant experience" (Stringer, 2004, p. 97) in a person's life. Once the epiphanies are identified, the researcher distinguishes the detailed features and then the elements that are associated with each epiphany. Each participant's experiences are then consolidated into a framework that shows his or her own personal perspectives and experiences in relation to the issue or problem. Once all of the individual reports have been completed, connections of similarities and differences are identified between the stakeholders (Stringer, 2004).

The categorizing and coding method is a traditional process where the large amount of data from each stakeholder's interview is divided into units of meaning. These units of meaning are then placed into the appropriate categories, subcategories, and themes that have been established by the researcher or participants involved. Organizing the information into a system of categories allows the participants to understand and interpret all the information and explains the significant features of the experiences appearing from the investigation (Stringer, 2004).

Analyzing the noninterview data is also an important part of the Think stage. Even though action research is mainly based on the participants' personal perspectives and experiences, noninterview data can present an extra resource to assist the participants in clarifying, supporting, and understanding the problem or issue being investigated (Stringer, 2004). Analyzing all the data then provides the background information necessary for effective action to be taken.

The final part of the cycle, the Act or Action stage, is divided into two processes. The first process is to inform all participants and stakeholders of the data that has been analyzed and the outcomes that have been established. This information can be presented in written reports, presentations, or performances. The research participants need to think carefully about how they will present their information so it will be communicated effectively to the appropriate audiences and so that it will represent understanding of their personal experiences and perspectives. Before any report, presentation, or performance is presented to an audience or focus group, member checking needs to be completed to ensure that the participants' experiences and perspectives are being presented accurately (Stringer, 2004).

The second part of this stage involves creating solutions to solve the problem or issue being investigated. All research participants need to be involved in the solution-solving process. The participants need to review all the information that has been presented to them in the reports, presentations, and performances, and they need to select the most important issues they would like to solve. The participants create a schedule outlining the action and order of events to be taken for each of the issues identified. This information is best recorded in a chart or table so that all information can be clearly read and reviewed by all participants. Once the action plan has been established and the tasks and steps are being implemented, each action needs to be assigned a supervisor. The role of this supervisor is to assist and support the people involved and monitor and communicate the progress to other participants. Regular meetings with the research participants should still occur at this stage so that the participants can report on their tasks and make any necessary changes and review the overall plan (Stringer, 2004).

Action research is powerful because of "its ability to allow researchers to tentatively state the problem, then refine and reframe the study by continuing iterations of the Look-Think-Act research cycle" (Stringer, 2004, p. 46). This means that the research can change as needed to meet the needs of the individual participants involved. It also allows the researcher to continue using these cycles until the problem or issue is completely solved. According to

Johnson and Christensen, "Action Research is a cyclical process because problems are rarely solved through one research study" (2000, p. 7).

OUTCOMES OF THE STUDY

Context of the Study

Blacksmith Primary is a double stream school, which is located in Hurstenville, Western Australia. It employs a variety of aged teachers with varying degrees of experience and understanding of the Curriculum Framework document. A new principal was appointed in Term 2 this year, so many changes are beginning to occur throughout the school.

This school has a strong community base with good relationships between the teachers, parents, and students of the school. The parents openly volunteer their assistance in all aspects of the school and often help out in many classrooms on a regular basis, and parents are always welcomed into the school environment.

Over the past two years at Blacksmith Primary there have been many changes implemented into the teaching of the new mathematics curriculum. During this time, the mathematics curriculum team has been investigating and identifying the key points and strategies needed to implement the new curriculum into the classrooms. This information has been taught to the classroom teachers through regular inservicing and collaborative learning sessions held during school staff meetings.

After a year of the teachers' being inserviced on the new curriculum, the mathematics team became interested in seeing how the classroom teachers, students, and parents were experiencing the new curriculum and how they felt it was working in each individual classroom. This came about after listening to many teachers talking to colleagues in the staff room. One teacher made the comment that "there is not enough time to fit the entire new mathematics curriculum into the school year." Another teacher said, "It is great how you can now integrate your math lessons into the themes that you are teaching." The mathematics team decided to start their investigation with the junior primary section of the school.

Student Experiences

Most of the students in the junior primary grades expressed a positive attitude to math in their classroom. Only one little boy had some concerns: "I don't always like math as sometimes it is too hard." All the students responded with answers that stated that they loved their teachers and were always able to gain assistance and help from their teachers when they required it.

During the individual interviews, each student gave a grand tour of his or her classroom, which they seemed to enjoy. They showed their workbooks, textbooks, the whiteboard ("this is where the teacher writes lots of sums," one student said), mathematics resources, and work hanging up around the room.

Most of the students did not make any comments about the textbook, but one girl said, "We don't use this book very much, I don't know why we even have it." When the students were showing the activities completed in their workbooks—most work was on worksheets that had been stuck into their book—one student said in a moaning voice, "We do lots of these worksheets, and they are so boring and easy. I complete them really quickly."

When the students came to the point in their tour that demonstrated the mathematics resources in their classrooms, most of the students started to play with the equipment. They displayed some of the things they liked to do and could do with it. All the students' responses were positive, and they expressed enjoyment with using the equipment, but they said they wished they could use it more, as "it is great fun playing with it, with my friends."

During the first focus group meeting, the students drew pictures of their favorite math lesson or activity that they really liked. All of the pictures drawn by the students involved using concrete material and working collaboratively with their friends. One girl drew herself and her friend playing on the computer, another drew herself playing with a group of children with the MAB blocks, and the boy drew himself and his friend measuring objects in the classroom.

At the second focus group meeting, the children created a poster with pictures of their likes and dislikes of mathematics. On the likes side, the children drew pictures of themselves playing games, using the mathematics equipment, using a calculator, working with their friends, and working on the computer. On the dislikes side, the children drew pictures of themselves sitting at their desks, the whiteboard, the mathematics textbook, and a math worksheet. When the students were asked why they disliked these things, they said, "They're boring and no fun because we can't play or talk to our friends."

Overall most of the students had a positive attitude to math but would like to see some changes so they can enjoy math more by regularly using concrete materials and learning collaboratively with their friends rather than working individually at their desks to complete worksheets or sums from the whiteboard.

Teacher Experiences

The teachers' responses all showed enthusiasm and enjoyment regarding teaching mathematics. All of the teachers had made some changes to the strategies and techniques they were using, but they still felt like they "have a lot more to learn." The teachers made many comments about their strong feelings about the use of textbooks, the mathematics resources, the children's attitudes, the need for continual inservicing, and time to share great ideas with their colleagues.

The teachers had both negative and positive responses to the use of the textbook. Some teachers liked "the textbook, as it makes for easy programming, and there is not much preparation for math lessons." However, the opinion of most teachers was that the textbooks were "repetitive and boring, and they don't allow for real-life problem solving." One teacher even stated, "The textbook is stopping me from completing creative and integrated lessons, and I hope we don't have to use them next year."

According to all the junior primary teachers there is a real concern with the mathematics resources in the school. All the teachers stressed the importance of using math equipment, "as it enhances the children's learning during a math lesson." However, all the teachers' responses stated that integrating math equipment into their lessons was difficult, as "the school is really lacking in resources" and "the resources are stored all over the school and you spend half your day trying to find them, and teachers don't have time for that." All the teachers would like to see more resources purchased. Some teachers felt that "more teacher reference books which show how to teach real-life problem solving skills to the students need to be purchased." Other teachers believed that "having a collection of websites that both the teachers and students could use would be very beneficial."

All of the teachers enjoyed seeing the students' enthusiasm for math lessons. Every teacher gave an example of a successful math lesson that he or she felt the students enjoyed and with which he or she believed they had real success. One teacher explained, "The children cheer every time I ask them to collect a calculator from the shelf." Another teacher explained how her class loved playing the Buzz game and asked if they could play it every day. All the teachers agreed that most math lessons could be "fun and enjoyable if they offered variety, interaction and the use of resources and equipment."

All of the teachers felt that they still required further inservicing with the mathematics curriculum. Many teachers felt that problem-solving skills and the Working Mathematically outcomes were two strong areas of concern. One teacher said, "I find it very difficult to assess the children in the Working Mathematically area as the pointers are not clear to me." Some teachers also stated that it would "be nice to take the time out to meet informally with other teachers and discuss different matters that arise in our classrooms or share successful math lessons with others."

Parent Experiences

All the parents expressed very strong and positive feelings toward the teachers at this school, and they congratulated them on the wonderful job they do. The parents believed that the teachers were "dedicated and enthusiastic teachers who showed love and kindness to all children in their classes." The parents felt the teachers explained new topics to the children clearly and concisely, and this was reflected in the children's being "able to complete their math homework independently, which is great for us."

The parents were very appreciative of the teachers for keeping them informed about their children's progress. One parent said, "It is great the way we receive feedback every term as you can always see how your children are doing." Another said, "I would like to see the school have parent teacher interviews at the end of Term One so we can see early on if our children are having any problems." The parents also found the teachers very approachable, and they liked how the teachers gave up their time whenever a parent requested it to discuss any concerns or questions they had regarding their children. One parent explained, "I have never had a teacher tell me they were too busy to see me, they always saw me that afternoon or the very next day."

All of the parents' school experiences were very different from what their children are experiencing today. One parent stated, "It is very different from when I went to school, we didn't play games or use equipment. The teacher just wrote on the blackboard and we copied it down." Another parent said, "All I remember is the constant math tests every Friday." The parents expressed concerns about their understanding of the teaching and assessing techniques being used in the classroom today. One parent said, "When the children bring home their portfolios there are all these tables on their work with ticks all over it. I don't read any of the writing, because I don't understand it. I just look at the marks to make sure that they are satisfactory or above." One parent suggested "holding information sessions to inform parents about the new teaching and assessing techniques."

Parent involvement and inclusion during math lessons in the classrooms was a strong point made by many of the parents. One parent believed "the children would learn better if there were more adults in the room," and another said, "More group work could be completed and the children could be grouped in ability like with reading." One parent expressed strongly that "having parents in the room is a great idea, but there has to be some time for the teacher to teach these children independently. Children also need breaks from their parents and time to spend with just their peers."

Students' Results From Standardized Tests

During one of the teacher focus group meetings, the teachers decided to review how the students were performing in the mathematics area. The teachers felt that this would give an extra perspective to the investigation. The students completed the standardized tests during Term 4 every year. The results from these tests show each child's strengths and weaknesses in the different mathematical strands. They also give individual class profiles that show the class's strengths and areas of concern. From the results, the teachers discovered that, since the last year, the students had really improved in the number and space strands. The Working Mathematically strand had not changed in most of the classes over the past two years, and in one class the results were very low.

Literature Study on Working Mathematically

Because there was a strong focus coming through from the participants' responses about needing to implement and come to a better understanding of problem-solving skills and the Working Mathematically outcomes, the teacher focus group decided to investigate what the literature had to say about these ideas. All the literature researched discusses the idea that the "process of Working Mathematically is learning to work like a Mathematician." This involves providing the children with problems to which they don't know the answers, and they have to play around and use a variety of strategies to solve them.

The teacher focus group also discovered a variety of schools around Australia that have changed their teaching styles and approaches to teaching mathematics to incorporate this new concept of working mathematically. The teacher focus group contacted these schools for advice and information to assist them in implementing new ideas into the Blacksmith Primary School community as part of the Action Agenda for 2005.

Action Agenda

After the three focus groups presented their group reports to each other, the parents and the teachers gathered together to form a new focus group. This larger group reviewed all the data collected in the group reports, the results from the standardized testing, and the information from the literature study. They used this information to identify the issues of concern and suggested solutions to accomplish the group's desired outcomes.

The group prioritized the issues and the areas of concern into the following categories: the school mathematical resources, the use of mathematics textbooks, parent involvement in the classrooms, teacher collaborative learning meetings, and inservicing the teachers on using real-life problem-solving techniques in their classrooms and gaining a better understanding of the Working Mathematically outcomes. The group decided that with Term 4 being such a short one, only some of the issues would be able to be addressed in this term and some would have to wait until Term 1, 2005.

Term 4, 2004

The teachers and parents felt that the school mathematical resources and the use of mathematics textbooks in the school required priority in Term 4. Throughout the reports, there was a strong emphasis on the importance of using resources and equipment in the teaching of the new mathematics curriculum. The school mathematical resources needed attention both with the reviewing of equipment in the school and with the purchasing of new equipment and teacher reference books. The teachers and students would also like to see the Internet being used more as a tool for teachers to gain ideas about teaching mathematics and for students to improve their mathematical skills.

To achieve these results the teacher and parent groups suggested the following action plan:

Complete a stock take of all the mathematics equipment in the school.

Discuss with the principal and the P&F Committee about including funds in next year's budget to purchase new mathematical resources and equipment next year.

Contact mathematical sales consultants for advice and samples of new resources and materials available to be purchased.

Create a property box on the school Web page for great mathematical Web pages to be included for easy access for all members of the school community.

Regarding the use of mathematics textbooks for next year, a meeting has been scheduled for November 15, 2004. At this meeting the principal and the junior primary teachers will discuss the positives and the negatives of using the textbooks and will then come to a decision about the approach for next year.

Term 1, 2005

In Term 1, 2005 the teachers will provide the parents with information and exploration sessions. These sessions will help the parents come to a clearer understanding of the teaching and assessing of mathematics today. At these sessions the parents will be given the opportunity to experience some hands-on problem-solving math lessons as well as gain greater understanding of the use of outcomes to teach and assess the students.

As all the participants expressed the belief that math lessons should be fun, a small committee of teachers, parents, and students will be established to create fun math sessions in which all junior primary can participate. The aim of these sessions is to expose the children to activities where they can work collaboratively to solve real-life problems and to include the parents in the classroom activities as requested.

Many of the teachers believed that teachers could learn a great deal from each other by just sitting and communicating together. This belief encouraged the teachers to look into having regular collaborative learning meetings for the junior primary teachers. At these meetings the teachers would gather together to gain advice and ideas from their colleagues as well as share successful math lessons.

As most of the teachers still feel that they lack expertise in the teaching of the new mathematics curriculum, a proposal will be put forward to the administration team of the school for a math consultant to come and inservice the teachers as part of the professional development plan for 2005. The information gathered from the literature study by the teacher focus group will assist in this area. The math consultant will inform the teachers of new strategies and techniques they will be able to use for teaching real-life problem-solving skills and the Working Mathematically outcomes to the students. The mathematics curriculum team would also investigate the purchasing of new teacher reference books with problem-solving and Working Mathematically teaching ideas.

CONCLUSION

I received the approval for the research project from the principal on August 24, 2004, and I presented the results to the principal and the teaching staff on Monday, November 22, 2004. The report was presented in the form of a big book that told the story of how the project was completed, the participants' experiences, and the actions that needed to be implemented. Even though the project was completed, there were still some areas of the action plan that will continue to be implemented this term and during Term 1, 2005. Now that I have completed a successful project, I feel more confident about completing action research projects in the future.

REFERENCES

Johnson, B., & Christensen, L. (2000). *Educational research: Quantitative and qualitative approaches.* Boston: Allyn & Bacon.

Spradley, J. (1979). *The ethnographic interview.* New York: Holt, Rinehart and Winston.

Stringer, E. (1999). *Action research* (2nd ed.). Thousand Oaks, CA: Sage.

Stringer, E. (2004). *Action research in education.* Upper Saddle River, NJ: Pearson.

APPENDIX B

Action Research Web Sites

The following Web sites are a small sample of those that now provide resources for people involved in action research activities and projects. Web sites continue to evolve, and some either disappear or are not updated, though the better sources provide ongoing access to useful information. To find Web sites with a particular focus, action researchers may use any of the popular search engines, entering key descriptors that will identify potentially relevant sources.

GENERAL

Action Research at Queen's University

An excellent resource, listing many links to action research Web sites providing information about action research programs, courses, conferences, publications, and student and faculty reports.

http://educ.queensu.ca/~ar

Action Research International—Online Journal

Refereed online journal of action research. Sponsored by the Institute of Workplace Research Learning and Development (WoRLD) at Southern Cross University. Editor Bob Dick.

www.scu.edu.au/schools/gcm/ar/ari/arihome.html

Action Research Network, Ireland

Provides a network of contacts for those involved in action research, action learning, and organizational learning in Ireland.

www.iol.ie/~rayo/

Action Research Resources

Provides an online journal, access to an online course, a number of useful papers, and other useful resources.

www.scu.edu.au/schools/gcm/ar/arhome.html

Action Research Resources

Action research resources on the Web—a collection of links sponsored by the PARnet, an interactive community on action research.

http://comm-org.wisc.edu/research.htm

ALARPM Public Home Page

Provides access to a wide range of resources, including links and forums for action research and action learning in education, health, management, land care, agriculture, and other areas.

www.alarpm.org.au

AR Expeditions—Action Research Journal

AR Expeditions is an online professional journal promoting a creative and critical dialogue between members of action research. It includes articles describing action research projects as well as strategies for conducting action research and hosts an online continuous dialogue about issues in action research.

www.montana.edu/arexpeditions/index.php

AROW—Action Research Open Web

Action Research Open Web provides open access to research, education, projects, and reports.

www2.fhs.usyd.edu.au/arow/

CARE

An important UK action research site located at the University of East Anglia that offers resources and links to other Web sites.

www.uea.ac.uk/care/

Collaborative Action Research Network (CARN)

A professional development network, focusing particularly on education.

www.did.stu.mmu.ac.uk/carn/whatis.shtml

Centre for Action Research in Professional Practice

Concerned with approaches to action research that integrate action and reflection and collaboration.

www.bath.ac.uk/carpp/

Collaborative Action Research Network

This Web site provides a variety of resources, in education, health, and welfare.

www.did.stu.mmu.ac.uk/carn/

emTech—Action Research

Action research resources—papers, sites, etc.

www.emtech.net/actionresearch.html

Jack Whitehead's Home Page

Offers a variety of resources including theses and links to other home pages.

http://people.bath.ac.uk/edsajw/

EDUCATION

Action Research

Rochester Teacher Center Teacher Research Program provides a variety of useful resources for teacher researchers.

www.rochesterteachercenter.com/teacherresearch.htm

Action Research Page

Provides a diverse selection of action research resources, including papers, links, and so on.

www.cudenver.edu/~mryder/itc/act_res.html

AR Special Interest Group

Action Research Special Interest Group of the American Educational Research Association.

http://coe.westga.edu/arsig/

An Action Research Approach to Curriculum Development

Action research has been used in many areas where an understanding of complex social situations has been sought in order to improve the quality of life.

http://informationr.net/ir/1-1/paper2.html

Classroom Action Research Overview

www.iusb.edu/~gmetteta/Classroom_Action_Research.html

Teacher Research (Action Research Resources)

Lists magazine and journal articles about action research, some of which are presented in full text form.

http://ucerc.edu/teacherresearch/teacherresearch.html

COMMUNITY DEVELOPMENT AND ORGANIZATIONS

Action Research for Environmental Management

Provides a summary of action research goals, differences between AR and mainstream science, and how reflection in action research is used to support change.

www.landcareresearch.co.nz/sal/ar_working.asp

Asset-Based Community Development Institute

At Northwestern University. Presents a strengths-based model of community development.

www.northwestern.edu/IPR/abcd.html

Informal, Adult, and Community Education

Presents an annotated bibliography about the nature and use of action research in informal education, adult education, and community education.

www.infed.org/research/b-actres.htm

Research for Action (RFA)

A nonprofit organization engaged in educational research and reform to improve educational opportunities and outcomes for all students. The site provides an extensive range of resources.

www.researchforaction.org/index.html

DISABILITIES

CAREERS & the disABLED Magazine

A career-guidance and recruitment magazine for people with disabilities.

www.eop.com/cd.html

Center for Computer Assistance to the Disabled

A non-profit organization offering help with assistive technology for computer access to the disabled.

www.axistive.com/1199/center-for-computer-assistance-to-the-disabled-c-cad-.html

Disabled People's International

A network of national organizations or assemblies of disabled people concerned with promoting the human rights of disabled people through full participation, equalization of opportunity, and development.

www.dpi.org

Disabilities-R-Us Chat Network

An Internet chat site for people with physical disabilities. Resources include a chat room, community message board, online disability meetings, disability links, and FAQs.

http://members.tripod.com/~disabled/

disABILITY Information and Resources

Lists Web sites of resources available to disabled people.

www.makoa.org/

HEALTH

CARAMASIA

Coordination of Action Research on AIDS and Mobility.

http://caramasia.gn.apc.org/

Ethnographic Action Research: UNESCO-CI

Ethnographic action research is an innovative research approach to poverty reduction.

http://portal.unesco.org/ci/en/ev.php-URL_ID=16619&URL_DO=DO_TOPIC&URL_SECTION=201.html

Health Education Library for People (HELP)

Provides health education resources for people in India, including resources for action research in holistic health.

www.healthlibrary.com

HIV InSite

Provides a comprehensive resource for information on HIV/AIDS treatment and prevention.

http://hivinsite.ucsf.edu

Poverty and Race Research Action Council

Seeks to link social science research to advocacy work to address problems of race and poverty. Health-related links—health; diet and nutrition; women, families and children—provide access to relevant articles, literature, and grants.

www.prrac.org/

SHARED (Scientists for Health and Research for Development)

Presents essential information on health research and development for developing countries—people, projects, organizations, electronic journals, other databases, networks, and groups.

www.shared-global.org/main.asp

Social Action for Health

A community-based organization that facilitates solution-focused changes in practice that lead to increased equity and better health for local people.

www.safh.org.uk/

Sustainable Health Action Research Programme (SHARP)

This Web site provides information about the Sustainable Health Action Research Programme (SHARP), an action research initiative focusing on community health development.

www.cmo.wales.gov.uk/content/work/sharp/index-e.htm

U.S. National Clearinghouse for Alcohol and Drug Information.

This site presents research briefs about the use of community action research in the prevention of alcohol and other drug problems.

http://ncadi.samhsa.gov/

The U.S. National Institute of Mental Health

Provides resources for practitioners, researchers, and the public. Includes information and research fact sheets, conference reports, research reports, funding opportunities, and patient education materials.

www.nimh.nih.gov/

World Health Organization

www.who.int/home-page/

PROJECT MANAGEMENT AND EVALUATION

Evaluation

A framework and tools for verifying achievement of project objectives.
www.worldbank.org/html/oed/evaluation/html/logframe.html

A Guide to the Project Management Body of Knowledge

Provides access to a useful text that will assist in managing complex action research projects.
www.pmi.org/publictn/download/2000welcome.htm

Outcome Mapping

Presents processes of evaluation for hard-to-measure projects, for example, community development. Includes sections on theory, frameworks, examples, projects, and publications.
www.idrc.ca/en/ev-26586-201-1-DO_TOPIC.html

WOMEN'S ISSUES

The British Columbia Centre of Excellence for Women's Health

Provides information about resources, publications, and grants related to women's health.
www.bccewh.bc.ca/

Feminist.com

Includes health and sexuality resources.
www.feminist.com/about/

Office of Women's Health

Promotes research, information, and service to women's health.
www.4woman.gov/owh/

The Women's Health and Action Research Centre (WHARC)

Promotes women's reproductive health in sub-Saharan Africa, conducting multidisciplinary and collaborative research, advocacy, and training on this issue. WHARC is a nongovernmental organization based in Benin City, Nigeria.
http://wharc.freehosting.net/

YOUTH

Institute for Community Research

The Youth Action Research Institute (YARI, formerly the National Teen Action Research Institute) promotes the use of action research for personal, group, and community development.
www.incommunityresearch.org/research/yari.htm

Reconnect Action Research Kit

A kit that shows how action research is applied to a youth program designed to reconnect at-risk young people to their families, schools, and the community.
www.facs.gov.au/internet/facsinternet.nsf/%20%20%20%20%20%20aboutfacs/programs/youth-reconnect_action_research_kit.htm

Youthwork Links and Ideas

Provides links to youth work resources, including advocacy, prevention, health, stories, programs, issues, and numerous other sites.
www.youthwork.com

Youth Work on the Internet

Acts as a virtual community for young people and professionals.
www.youth.org.uk

Youth Worker Resources

Tools for youth work.
www.youthworker.org.uk/

REFERENCES

Alinksy, S. (1971). *Rules for radicals*. New York: Random House.

Anderson, G., Herr, K., & Nihlen, A. (1994). *Studying your own school: An educator's guide to qualitative practitioner research.* Thousand Oaks, CA: Corwin.

Atweh, B., Weeks, P., & Kemmis, S. (Eds.). (1998). *Action research in practice: Partnerships for social justice in education.* New York: Routledge.

Benner, P. (1994). *Interpretive phenomenology: Embodiment, caring, and ethics in health and illness.* Thousand Oaks, CA: Sage Publications.

Berger Gluck, S., & Patai, D. (Eds.). (1991). *Women's words: The feminist practice of oral history.* New York: Routledge.

Black, J. P. (1991). *Development in theory and practice: Bridging the gap.* Boulder, CO: Westview.

Block, P. (1990). *The empowered manager: Positive political skills at work.* San Francisco: Jossey-Bass.

Brown, D. (1970). *Bury my heart at Wounded Knee: An Indian history of the American west.* London: Vintage.

Bruner, J. (1990). *Acts of meaning.* Cambridge, MA: Harvard University Press.

Burnaford, G., Fischer, J., & Hobson, D. (Eds.). (1996). *Teachers doing research: Practical possibilities.* Mahwah, NJ: Lawrence Erlbaum.

Calhoun, E. (1993, October). Action research: Three approaches. *Educational Leadership, 51*(2), 62–65.

Christians, C., Ferre, J., & Fackler, M. (1993). *Good news: Social ethics and the press.* New York: Oxford University Press.

Collins, P. (1991). *Black feminist thought.* New York: Routledge.

Coover, V., Deacon, E., Esser, C., & Moore, C. (1985). *Resource manual for a living revolution.* Philadelphia: New Society.

Coughlan, D., & Brannick, T. (2004). *Doing action research in your own organization* (2nd ed.). Thousand Oaks, CA: Sage Publications.

Denzin, N. K. (1989). *Interpretive interactionism.* Newbury Park, CA: Sage Publications.

Denzin, N. K. (1997). *Interpretive ethnography.* Thousand Oaks, CA: Sage Publications.

Denzin, N. K., & Lincoln, Y. S. (Eds.). (1994). *Handbook of qualitative research.* Thousand Oaks, CA: Sage Publications.

Denzin, N. K., & Lincoln, Y. S. (Eds.). (2005). *Handbook of qualitative research* (3rd ed.). Thousand Oaks, CA: Sage Publications.

Derrida, J. (1976). *Of grammatology.* Baltimore: Johns Hopkins University Press.

Deshler, D. (1990). Conceptual mapping: Drawing charts of the mind. In J. Mezirow & Associates (Eds.), *Fostering critical reflection in adulthood* (pp. 336–353). San Francisco: Jossey-Bass.

Fals-Borda, O., & Rahman, M. (Eds.). (1991). *Action and knowledge: Breaking the monopoly with participatory action-research.* New York: Apex.

Fish, S. (1980). *Is there a text in this class? The authority of interpretive communities.* Cambridge, MA: Harvard University Press.

Foucault, M. (1972). *The archaeology of knowledge.* New York: Random House.

Foucault, M. (1984). *The Foucault reader* (P. Rabinow, Ed.). Harmondsworth, UK: Penguin.

Goodenough, W. (1963). *Cooperation in change: An anthropological approach to community development.* New York: Russell Sage.

Greenwood, D., & Levin, M. (2006). *Introduction to action research: Social research for social change* (2nd ed.). Thousand Oaks, CA: Sage Publications.

Guba, E. G. (1990). *The paradigm dialogue.* Newbury Park, CA: Sage Publications.

Guba, E. G., & Lincoln, Y. S. (1989). *Fourth generation evaluation.* Newbury Park, CA: Sage Publications.

Gustavsen, B. (2001). Theory and practice: The mediating discourse. In P. Reason & H. Bradbury (Eds.), *Handbook of action research* (pp. 17–26). Thousand Oaks, CA: Sage Publications.

Habermas, J. (1979). *Communication and the evolution of society* (T. McCarthy, Trans.). Boston: Beacon.

Habermas, J. (1984). *The theory of communicative action: Reason and the rationality of society.* Boston: Beacon Press.

Haebich, A. (1988). *For their own good.* Perth: University of Western Australia Press.

Hart, E., & Bond, M. (1995). *Action research for health and social care: A guide to practice.* London: Open University Press.

Hendricks, C. (2006). *Improving schools through action research: A comprehensive guide for educators.* Boston: Allyn & Bacon.

Heron, J. (1997). *Co-operative inquiry: Research into the human condition.* Thousand Oaks, CA: Sage Publications.

Holly, M. L., Arhar, J., & Kasten, W. (2004). *Action research for teachers: Travelling the yellow brick road* (2nd ed.). Upper Saddle River, NJ: Prentice Hall.

Huyssens, A. (1986). *After the great divide: Modernism, mass culture, postmodernism.* Bloomington: Indiana University Press.

Kelly, A., & Gluck, R. (1979). *Northern Territory community development.* Unpublished manuscript, University of Queensland, St. Lucia, Australia.

Kelly, A., & Sewell, S. (1988). *With head, heart, and hand.* Brisbane, Queensland, Australia: Boolarong.

Kemmis, S., & McTaggart, R. (1999). *The action research planner* (3rd ed.). Geelong, Australia: Deakin University Press.

Kickett, D., McCauley, D., & Stringer, E. (1986). *Community development processes: An introductory handbook.* Perth, Western Australia: Curtin University of Technology.

Kincheloe, J. (1991). *Teachers as researchers: Qualitative inquiry as a path to empowerment.* London: Falmer.

Koch, T., & Kralich, D. (2006). *Participatory action research in healthcare.* London: Blackwell.

Levin, M., & Greenwood, D. (2001). Action research and the struggle to transform universities into learning communities. In P. Reason & H. Bradbury (Eds.), *Handbook of action research: Participative inquiry and practice* (pp. 103–113). Thousand Oaks, CA: Sage Publications.

Lewin, K. (1946). Action research and minority problems. *Journal of Social Issues, 2,* 34–46.

Lincoln, Y. S. (1995). Emerging criteria for quality in qualitative and interpretive inquiry. *Qualitative Inquiry, 1,* 275–289.

Lincoln, Y. S., & Guba, E. G. (1985). *Naturalistic inquiry.* Beverly Hills, CA: Sage Publications.

Lincoln, Y. S., & Tierney, W. (1996). *Representation and the text: Reframing the narrative voice.* Albany: SUNY Press.

Lyotard, J.-F. (1984). *The postmodern condition: A report on knowledge.* Minneapolis: University of Minnesota Press.

Malinowski, B. (1961). *Argonauts of the western Pacific: An account of native enterprise and adventure in the archipelagoes of Melanesian New Guinea.* New York: E. P. Dutton. (Original work published 1922)

McCauley, D. (1985). *Community development.* Unpublished manuscript, Perth, Western Australia.

McMurray, A., & Pace, R. W. (2006). *Action research in learning organizations.* Thousand Oaks, CA: Sage Publications.

McNiff, J., Lomax, P., & Whitehead, J. (1996). *You and your action research project.* Bournemouth, UK: Hyde.

McNiff, J., & Whitehead, J. (2006). *Action research for teachers: A practical guide.* Abingdon, UK: David Fulton.

McTaggart, R. (Ed.). (1997). *Participatory action research: International contexts and consequences.* Albany: SUNY Press.

Mead, G. H. (1934). *Mind, self, and society: From the standpoint of a social behaviorist.* Chicago: University of Chicago Press.

Milgram, S. (1963). Behavioral study of obedience. *Journal of Abnormal and Social Psychology, 67,* 371–378.

Mills, G. (2006). *Action research: A guide for the teacher researcher* (3rd ed.). Columbus, OH: Merrill/Prentice Hall.

Morton-Cooper, A. (2000). *Action research in healthcare.* Oxford: Blackwell Publications.

Neumann, A., & Peterson, P. (1997). *Women, research, and autobiography in education.* New York: Teachers College Press.

Noffke, S., & Stevenson, R. (Eds.). (1995). *Educational action research: Becoming practically critical.* New York: Teachers College Press.

Peck, M. S. (1993). *A world waiting to be born: Civility rediscovered.* New York: Bantam.

Reason, P. (Ed.). (1988). *Human inquiry in action: Developments in new paradigm research.* Newbury Park, CA: Sage Publications.

Reason, P., & Bradbury, H. (Eds.). (2001). *Handbook of action research.* Thousand Oaks, CA: Sage Publications.

Reason, P., & Bradbury, H. (Eds.). (2007). *Handbook of action research* (2nd ed.). Thousand Oaks, CA: Sage Publications.

Reeb, R. (2007). *Community action research: Benefits to community members and service providers.* Binghamton, NY: Haworth Press.

Sagor, R. (2004). *The action research guidebook: A four-step process for educators and school teams.* Thousand Oaks, CA: Corwin Press.

Schmuck, R. (2006). *Practical action research for change* (2nd ed.). Boston: Allyn & Bacon.

Schouten, D., & Watling, R. (1997). *Media action projects: A model for integrating video in project-based education, training and community development.* Nottingham, UK: University of Nottingham Urban Programme Research Group.

Senge, P., Kleiner, A., Roberts, C., Ross, R., & Smith, B. (1994). *The fifth discipline fieldbook.* Garden City, NY: Doubleday.

Smith, D. E. (1989). Sociological theory: Methods of writing patriarchy. In R. A. Wallace (Ed.), *Feminism and sociological theory* (pp. 34–64). Newbury Park, CA: Sage Publications.

Spradley, J. P. (1979). *The ethnographic interview.* New York: Holt, Rinehart & Winston.

Spradley, J., & McCurdy, D. (1972). *The cultural experience: Ethnography in complex society.* Prospect Heights, IL: Waveland Press.

Stake, R. (1994). Case studies. In N. Denzin & Y. Lincoln (Eds.), *Handbook of qualitative inquiry* (pp. 236–247). Thousand Oaks, CA: Sage Publications.

Stringer, E. (2002). *Parent-teacher associations in East Timor.* Dili, East Timor: Ministry for Education, Culture, Youth and Sports/UNICEF.

Stringer, E. (2004). *Action research in education.* Upper Saddle River, NJ: Pearson.

Stringer, E. (2007). "This is so democratic!" Action research and policy development in East Timor. In P. Reason & H. Bradbury (Eds.), *Handbook of action research.* Thousand Oaks, CA: Sage Publications.

Stringer, E., & Dwyer, R. (2005). *Action research in human services.* Upper Saddle River, NJ: Pearson.

Stringer, E., & Genat, W. (2004). *Action research in health.* Upper Saddle River, NJ: Pearson.

Tillman, L. (2002). Culturally sensitive research approaches: An African American perspective. *Educational Researcher, 31*(9), 3–12.

Trinh, M.-H. (1992). *Framer framed.* New York: Routledge.

Tyler, S. (2006). *Communities, livelihoods, and natural resources: Action research and policy change in Asia.* Rugby, UK: Intermediate Technology Development Group Publishing.

Unger, R. (1987). *Social theory: Its situation and its task.* Cambridge: Cambridge University Press.

Van Willigen, J. (1993). *Applied anthropology.* South Hadley, MA: Bergin & Garvey.

Vlachos, E. (1975). *Social impact assessment: An overview.* Fort Belvoir, VA: Army Corps of Engineers, Institute of Water Resources.

Wadsworth, Y. (1997). *Everyday evaluation on the run* (2nd ed.). St. Leonards, New South Wales, Australia: Allen & Unwin.

West, C. (1989). *The American evasion of philosophy.* Madison: University of Wisconsin Press.

Whitehead, J., & McNiff, J. (2006). *Action research: A living theory.* Thousand Oaks, CA: Sage Publications.

Whyte, W. F. (1984). *Learning from the field: A guide from experience.* Beverly Hills, CA: Sage Publications.

Winter, R., & Munn-Giddings, C. (2001). *A handbook for action research in health and social care.* London: Routledge.

Witherall, C., & Noddings, N. (Eds.). (1991). *Stories lives tell: Narrative and dialogue in education.* New York: Teachers College Press.

INDEX

ABOUT THE AUTHOR

Ernie Stringer

After an early career as primary teacher and school principal, Ernie was a lecturer in teacher education for a number of years at Curtin University of Technology in Western Australia. From the mid-1980s he worked collaboratively with Aboriginal staff and members of the community to develop a wide variety of innovative and highly successful education and community development programs at Curtin's Centre for Aboriginal Studies. Consultative services with government departments, community-based agencies, business corporations, and local governments helped many people work more effectively in Aboriginal contexts. In recent years, as visiting professor at universities in New Mexico and Texas, he taught action research to graduate students and engaged in projects with African American, Hispanic, and other community and neighborhood groups. As a UNICEF consultant from 2002 to 2005 he engaged in a major project that assisted development of schools in East Timor. He is author of *Action Research* (1999), *Action Research in Education* (2004), *Action Research in Health* (with Bill Genat, 2004), and *Action Research in Human Services* (with Rosalie Dwyer, 2005). He is a member of the editorial board of the *Action Research Journal* and is president of the Action Learning and Action Research Association (ALARA).

He can be reached by email: e.stringer@exchange.curtin.edu.au or erniestringer@hotmail.com.